Beautiful Beasts, Beautiful Lands

Mr Kasasira with his beautiful cow, Byeyera

Beautiful Beasts, Beautiful Lands

The fall and rise of an African national park

Mark Infield

Whittles Publishing

Whittles Publishing Ltd.,
Dunbeath,
Caithness, KW6 6EG,
Scotland, UK

www.whittlespublishing.com

Printed and bound by CPI Group (UK) Ltd, Croydon, CR0 4YY

This book is printed using paper from well-managed
forests and other controlled sources.

Contents

For Moses Turyaho, whose love and dedication to Mburo was
beyond measure

Preface

I always felt that working in nature would suit me. It would be fulfilling and inspiring. I was right on that score. I thought it would be relaxing and undemanding. There I was wrong. The story I tell here describes lessons learned from practical experience that I hope provide a different perspective to the mainstream of modern conservation. The ideas in it are not my invention and have been espoused by many. Nonetheless, the practical employment of cultural values in conservation, especially the values that connect people to place, is surprisingly rare. My critique of the scientific and economic perspectives that dominate modern conservation does not invalidate them or diminish their roles but seeks to challenge them and to ask for a more nuanced assessment of their implications for conservation.

My conclusion is that we must protect nature for and through its worth; and that its worth does not lie exclusively or even primarily in the values of conservationists, scientists, or economists, but in the values of those who live with nature. To reach this conclusion I have had to wrangle with words used all the time without difficulty but which express ideas without simple or single meanings. The word 'culture' is widely used and broadly understood. But when applied to a practical problem it is so all-encompassing that I hardly know where to start or end with it. Whether I try to grasp it tightly as a tool, or I open my hand to examine it more closely, it runs through my fingers like fine sand.

'Community', the entire context for conservation in truth, is another word known well and easily used. But it is a mysterious and subtle concept, loaded with connotations of solidarity, supportiveness, interconnectivity, public interest, and all good things, however unlikely. Not, therefore, something to easily deal with, work with or depend on. 'Nature', a word without which I could not have begun to write or think about this book is yet a word without meaning for most people on earth as they make no separation between human life and all other life.

'Beauty' lies at the core of this the book but is so particular that it is almost impossible to translate across cultures. I once spent three days asking residents of the Subak rice terraces of Bali whether they considered their famously beautiful landscape to be beautiful. To ask the question we described the terraces as "rice growing in a place where all that should be there is there, with nothing that should not". The place was so extraordinarily beautiful to me that I had to bite my tongue at such a description.

The idea of 'land' is central to this book. Land is the terrain under our feet but also the place we go to in our thoughts. Land is where we live, the history of its people and peoples, the uses we put it to, the institutions that manage it and the spirits that inhabit it. Land is connection, our motherland, fatherland, homeland, heartland, even our home water.

In the end this book is about the love of nature and how stories can help us understand something about conservation, see its contradictions and confusions, recognize its successes and failures, and perhaps, point to a direction for its future.

Mark Infield

Foreword

As I write this I am immersed in the burgeoning spring of rural Oxfordshire – the air is scented with an intoxicating mix of bluebell, horse chestnut and hawthorn blossom, the woods glow in an aquamarine haze of bluebells, with steel-grey beech trunks and fresh soft lime-green leaves glowing above, noisy with birdsong and slow-buzzing insects. Everywhere here there is new life; the sap surging in the plants is almost tangible, and I can't help but feel my spirits rise too.

I am almost certain that the author of this wonderful, inspiring book, Mark Infield, is feeling the same sensory cornucopia in his home in Sussex, and is, like me, sucking in every second of the UK spring to replenish his winter-tired body and soul. I know this because we come from very similar cultural, geographical, educational roots, which deeply influence our sense of beauty and our connection to nature. But this rootedness in the seasonal cycles of southern England is not universally shared; one person's bluebell dream is another's low-diversity woodland. Abstractly, these differences in view about what makes for 'good' nature are well rehearsed in the conservation literature, but in practice our own deep connections with (or even prejudices about) nature remain unspoken and implicit. And when people who are wired to think about nature in one way through their own experiences wield power to change whole landscapes in a completely different context (such as the British in East Africa), terrible outcomes can ensue.

This book is about beauty and love for nature, but also about human connections and empathy. It's about the twisted history of conservation in which high ideals meet dirty reality – and do not come well out of the encounter. It shows with rare, refreshing, honesty how a naive young person wanting adventure can, despite their doubts, be swept along in the rapids of prevailing thought. It is one of the best, most honest and inspiring books about conservation I have ever read.

Mark Infield traces a thread from his childhood in Ashdown Forest, UK, to Mburo, a relatively unknown place in Uganda, and thence to his career in conservation and his reflections on where we are going wrong in the way we think about nature and our relationships with it – and with other people who may have a completely different worldview from ours. He shows the vital importance of taking the long view – including understanding the deep roots that can culminate in disastrous outcomes, and how surface impressions of current landscapes and livelihoods may be unrepresentative of the past. For example, the devastating impact of rinderpest on livestock and wildlife in late 19th-century East Africa led to societal calamity and created the impression of vast empty lands that conveniently aligned with romantic ideals of wilderness among colonial powers, facilitating the creation of exclusionary nature reserves.

Deep drivers of people's relationships with the land, with each other and with institutions can lead to complex, conflictual situations into which, as Mark shows, the foreign conservationist can blunder. That foreign conservationist can then sit in the awkward position of being both a powerful player with external funding and connections and a puppet for vested interests. The temptation is to propose and enact simplistic 'solutions' rather than sitting down together with the people who actually use the land, with humility, to understand both the situation and (importantly) what role they may usefully play in improving matters. Mark's description of his long involvement with Mburo is inspiring in its focus on friendship and teamwork with his Ugandan colleagues, in its honesty about the challenges he faced and where he fitted in (as the tame *muzungu* who could demonstrate that the world cared about what happened there, and could bring in money and influence, as well as ideas).

Mark has a strong focus on the need to start by understanding how people's cultural connections to nature affect their views and responses. I particularly enjoyed the 'through the looking glass' feeling of slight dislocation and half-glimpsed understanding when Mark explored how the Bahima people see Mburo, their conception of beauty and the land's 'rightness' for their cattle. How they conceptualise beauty is so different from the cultural framing of southern England that even finding the language to express it is hard. But we don't need to fully understand how others relate to an area in order to get to the essence of what we, as conservationists, need to do. As Mark says, 'Although our own decisions and sacrifices are motivated by love, not lucre, we seem unable to believe that others might be the same.'

This book should be required reading for all conservationists, particularly those just starting out in their careers. For me, it underscores yet again the need for those of us who meddle in others' homelands to have some humility. As Mark says, 'I cannot easily set aside the dualism that lies at the heart of my culture, but fortunately I don't need to do that. I just need to place my own beliefs no higher and no lower than the myriad ways in which others experience their universe.'

Professor EJ Milner-Gulland

Tasso Leventis Professor of Biodiversity; Director of the
Interdisciplinary Centre for Conservation Science, University of Oxford

Acknowledgements

This story opens with my arrival in Uganda in 1981. Many people have been part of it since then and I owe them all thanks. Rob Malpas introduced me to conservation when he took me to help him with his work in Uganda. John Hanks and many others at the Institute of Natural Resources helped me explore links between communities and conservation. Fredrick Sabela taught me how to behave.

I was guided during the first years at Mburo by Moses Turyaho and Arthur Mugisha. Eric Edroma and Fred Kayanja at Uganda National Parks encouraged me along the way. Many people played critical roles in the Mburo project including Agrippinah Namara, who led the community surveys, and Florence Kabwachera and William Mwesigye, Uganda's first Community Rangers. Mark Stanley Price and Ed Barrow mentored me during its implementation.

At Fauna & Flora International Jon Hutton, Mark Rose, Ros Aveling and Abi Entwistle are due thanks for supporting my work on culture in conservation, and Katie Frohardt for being its greatest champion. The MacArthur Foundation funded the cultural values project for 10 years thanks to Michael Wright and Elizabeth Chaudri's encouragement.

For accepting me as his PhD student, for teaching me what research was and for pushing me when pushing was needed I owe a special thanks to Michael Stocking.

I owe more than I can say to the advice, support and friendship I received from Charles Muchunguzi and Patrick Rubagyema. They carried the job title of 'Research Assistant' but without their knowledge and commitment my research would have failed. And without the patience and tolerance of the Ranch 10 community, especially Kyabahindi who accepted me into his *Eka*, this book would not have been written.

Bill Adams advised me on the kind of book I might write and how to begin it and helped in every way. Mike Unwin and EJ Milner-Gulland were kind enough to read drafts and provide comments and encouragement. Keith Whittles and the team at Whittles Publishing were very patient and guided me through the process of publication. Working with my editor, Caroline Petherick – the Wordsmith – was a privilege and a pleasure.

I would like to thank Sandy for her unflagging patience and encouragement and for never allowing me to give up on the project. My father read the very first draft and praising it unstintingly. And finally, I thank Arthur Mugisha for a friendship and partnership that has lasted 35 years and for his wisdom, warmth and dedication to conservation that has inspired me since we first met.

Introduction

In 2022 the first African Protected Areas Congress was held in Kigali, Rwanda. Africa's first national park had been established in 1925 by colonial authorities, just across the border in what was then the Belgian Congo, now the Democratic Republic of Congo. That park established the dominant exclusionary model of protected areas in Africa, and it set a tone for their creation and management that would endure for a century. The participants at the 2022 congress, however, especially the many voices of indigenous and local peoples, gave a clear call to develop an African vision of conservation that put people back at its centre.

Demands to refute, reform or abandon the 'fines and fences' approaches to protected areas are not new. This book draws on the history of Lake Mburo[1] National Park in Uganda to describe the efforts made over the last 40 years to create a more open and inclusive approach to nature conservation. I investigate the nature of nature, and the nature of efforts to protect it, focusing on Lake Mburo National Park and on the years between 1990 and 2015, when I worked there. I also reflect on the modern conservation endeavour through the lens of my relationship with nature. I spent 20 years in Africa, mostly in Uganda, but I also worked in South Africa and Cameroon. I lived and worked in Vietnam and Indonesia for ten years and have been lucky enough to visit many countries and protected areas. All these experiences have informed my ideas about nature and how best to protect it.

Efforts to protect nature and wild lands sometimes seem impossibly complicated. Perhaps the first exponents of modern conservation, who agitated so successfully in the 19th century to protect natural areas from development or tropical birds from the hat makers, thought the same. But they persisted. Today, more than ever before, and despite the difficulties, we too must persevere. Threats to our natural world are driven by complex, interlinked global phenomena whose impacts are becoming more and more evident. Industrial production, fuelled by the escalating expectations of wealth amongst the

---

1 This name (and others like it, starting with M and a consonant) is pronounced as in "Mmm, delicious".

already wealthy and the growing demands for necessities from less affluent populations, continues to consume and pollute the planet. Expanding agriculture, both industrial and subsistence, fragments the land and degrades ecosystems. Unsustainable levels of consumption weaken or destroy natural resources and threaten species. And the spectre of the human-induced climate emergency hangs over it all. Together, these destructive forces have driven the shocking declines in animal and plant life over the past half-century that we have become familiar with from news reports and stories but which most of us barely notice.[2]

Notwithstanding the shock of the Covid 19 pandemic in 2020, social and economic development across countries and continents is happening with a level of intensity and immediacy unimaginable even a few decades ago. Startling new technologies seem to appear daily. Communications and the flow of information never sleep. Financial trading works at the speed of light according to decisions made by algorithms. All this makes for a world in which it is hard to find solutions to conservation problems, hard to implement them, and hard to determine if they have been of any use by the time the next set of problems arrives. At the same time the hegemony of the west and its values, philosophies and understandings of the world are increasingly contested, and positive representations of its historical global role challenged.

This book being about just one thing, I thought it would be simple to write, and I hoped it would be simple to read. But difficult questions arise from any discussion of conservation, and they cannot be avoided if we hope to succeed better in the future than we have to date. Though the question may be seen as a banal one, we still need to ask what nature is before we can answer how best to protect it.

Nature transforms the physical elements of the planet from inert materials into something elevated with the almost magical qualities of life that suffuse all organisms. Some people distinguish nature from the not natural, and confine their view of nature to what they perceive as alive. Others incorporate the elements – fire, wind and water, the earth below and the sky above – into their understandings of nature. Yet others, including my colleagues and teachers, define nature by its names and characteristics, and indeed the act of describing and naming species lies at the heart of conservation science. For many, however – the majority perhaps – the question 'what is nature?' has no meaning, as the

---

2 Shifting Baseline Syndrome describes why we fail to recognise even dramatic changes over time. What we experience tell us what is normal, but not that this normal is ever shifting. We struggle as individuals and societies to recognize gradual, incremental change.

   One sunny summer day my daughter and I were watching a butterfly in the garden. I told her that when I was a boy, I used to see lots of butterflies. She looked at me in some surprise and assured me happily that she also saw lots – she had seen three or four the day before, at least three the previous weekend, and even more before that. She looked dubious when I told her that I would see butterflies in their tens and even hundreds in my mother's garden.

   But it was not until I travelled to Africa that I experienced what pre-industrial numbers of butterflies look like; clouds, drifts and deluges of butterflies, carpets of butterflies. We all grow to be content, it seems, with what we know and experience. This fact raises serious questions about the pursuit of conservation on behalf of future generations, who may be entirely satisfied with a diminished natural world even if we are not.

source of the question, the human mind, is not separate from the subject. How can we ask the nature of something we are ourselves part of and indivisible from?

These questions, though puzzling, don't require you and me to generate answers to them for us to embrace the challenge of protecting nature. All that is needed is that whatever the answers, we accept them all. From Aristotle to today's ecologists, there have been interpretations of nature through the lens of science and attempts at empirical observation, but there are as many other ways of understanding nature as there are people in the world.

Philosophers, psychologists and anthropologists have all done their best to say what nature is, what it means to humanity, and why it should be protected. Even economists have got in on the act. Religions, faiths and indigenous cosmologies and beliefs do the same. I make no attempt to propose my own or critique others. Instead, I describe how, through the course of my work and my life, my thinking about nature and the pursuit of conservation has evolved. Some insights, some understandings and some suggestions have emerged, but no answers to the big questions. I have come to think that what nature is depends as much on what you think as on what I think. And what you think depends, ultimately, on the interaction between 'culture' and lived experience. This book, it has turned out, could not be just about nature. It had to be about culture too, and how nature and culture fuse in our relationships with the world around us.

The thing about culture is that it is a trickier idea than it is a word. We are familiar with 'culture' as a word; we use it all the time, and quite happily. But when we try and work with the idea of culture, its meaning seems to run through our fingers like sand – the harder we grasp it, the less we see of it. So it is only in an open hand that we can hold it and see it. This story is, as much as anything, about how the values of nature – its worth as I experienced it, and the emotions it engendered in me – are what have driven my wish to protect it and to inspire others to do the same. This recognition has eventually directed me to try to respond to the values of the natural world as others perceive them, to be guided by how others understand the worth and meaning of nature, and to validate the feelings of others as well as my own.

This on the face of it seems a very little thing. And indeed, I feel it ought to be thought of as a very little thing. I set out to write this book, though, because my experience suggests that modern conservation sees things in a different light.

My journey away from the fundamental assumption that nature and humanity had to be separated to achieve conservation began with recognition of my own failures and my concerns over the unexpected consequences of my actions. I describe my part in the unjust creation of Lake Mburo National Park, its decline, and the subsequent struggle, which I was part of, to rescue it. This experience led me to conclude that modern conservation had wandered down a narrow path dominated by arguments informed by materialism and economics. I believe we must turn around and embrace the deeper worth of nature that inspired modern conservation in the 19th century and base our efforts to protect nature on this.

At the age of ten, sitting at my desk in the prefab cabin that served as our classroom, smelling of wood resin and chalk, I wrote a poem. Our teacher had marking to do and wanted us to get on with something quietly. She talked to us briefly about poems, and then asked us all to write one. Heads bowed in concentration broken by the occasional scuffle or giggle, we tried.

My father's head is filled with poems. He can recite any number to suit any occasion. One of his favourites is 'Pied Beauty' by the poet priest, Gerald Manley Hopkins. Over years his impromptu recitations fixed it in my mind, too:

> Glory be to God for dappled things –
> For skies of couple-colour as a brinded cow;
> For rose-moles all in stipple upon trout that swim.

It is no surprise, then, that when we were asked to write a poem, mine was about nature. I was quickly finished – ten or twelve short lines – and I presented them with the unaffected confidence of a child.

I no longer have that poem, written in pencil in an exercise book I had covered with a photo of Apollo astronauts in their capsule, my name stencilled in a corner – but I remember it, and how it was received by my teacher. Most afternoons, summer and winter, sunny or rainy, I walked from our house with my dog along the edge of the field that swept up the small valley to the woods above, hummocked with holes and hills and banks. The smaller holes were no more than shallow scrapes filled with drifting leaves in autumn, but the larger ones held ponds, dark with overhanging trees, fringed with reeds and brambles, hosting moorhens and on occasion dark-headed mallard ducks.

The largest had a silver birch growing from an island of damp earth no more than a couple of feet square, and just a yard from the plashy pond edge. It was my habit to jump the narrow gap over the tea-brown water, turn, and with my back against the trunk, stand quietly there. Little would happen. Nothing would happen. Nor was I waiting for anything. A squirrel might run along a branch above me. One of the moorhens might scull slowly across the pond – I would turn to watch it go. My dog covered vast circles around me, always moving as I stood unmoving, tracing the scents of rabbit or fox or badger.

My still silent watch combined sensations of alert attention and deep peacefulness. I anticipated my response to a bird landing nearby or a small creature passing. I would settle into a deep contentment, a passive meditation, a reflection on the water gently moving around my island, the pale lemon-green chevrons of spring birch leaves gently moving in the breeze, or the delicate tracery of fine branches against a purple winter sky.

This is what I wrote about. The sounds and smells and sights of this little woodland and its pond, the sweep and flow of fields and hedges and the small seasonal streams that marked their margins were the world of fact and imagination that I inhabited.

These ordinary certainties, individual and collective, had permeated and gathered in my core.

I was expressing something particular and personal about nature that sleepy afternoon in the quiet classroom. My solitary watches in the woods must have been very much with me then, as they are with me now as I write.

I remember the teacher looking at me with a quizzical expression. After a long moment, my happy confidence draining away, she asked, 'Did someone help you with this?' It was a strange question. It communicated disquiet, that there was something a little peculiar in what I had written. With another doubtful look she sent me back to my desk to write another poem. I wrote about running a race. I liked to run. I could run fast. It seemed a safer subject.

The teacher's reaction to that first poem taught me that my feelings about nature were not a shared experience, not something everyone would relate to, but something of my own. I discovered that that world was best kept to myself. I learned to hide my feelings about nature, or more accurately, the feelings that nature gave me.

My experiences of nature fused with my imaginings of nature. From the books my parents read I learned that wildebeests and crowned cranes lived on the veldt, squirrels slept soundly through the winter on piles of acorns, and badgers lived in the wild woods with the stoats and weasels. Rudyard Kipling's *Just So Stories* gave me richly imagined landscapes full of creatures, half-real, half-mythical. Kangaroos and dingoes ran forever through parched Australian copses of gum trees; rhinoceroses rubbed themselves against palm trees; and camels wandered grumpily through the sand dunes of a great desert. Elephants, crocodiles and rock pythons inhabited the banks of the 'great grey-green, greasy Limpopo River, all set about with fever-trees'.

I knew nothing of real elephants or pythons, but henceforth associated them with a vast river moving heavily between banks dotted with luminous trees – an image I found to be surprisingly close to the real thing when I eventually saw the real thing. Perhaps, though, the book that had the most powerful grip on my imagination was *The Jungle Book*, the story of a boy living in the jungle, the brother of wolves and bears. Mowgli was a boy my age, who grew older as I grew older, a boy who lived in community with nature. I wanted it too.

The magic of these fictional worlds with their almost real animals and birds lent magic to the world I inhabited. The fields, hedgerows and woodlands I wandered harboured, I was sure, creatures just as magical, and the difficulty of seeing wildlife in the English countryside – there isn't that much of it in the first place and what little there is tends to be small and hidden – gave space for me to create my own land thronging with wonderful creatures. It is perhaps here that my conservation journey started. The real world was joined with the imagined, and the ordinary became the extraordinary. The physical world of nature that I would learn to describe in the dispassionate terms of scientific observation was one with the experienced world of nature that I could only describe in terms of the individual and the personal. I could comprehend nature only in my own terms.

My feelings for the landscape in which I lived were not the product of an ancient tradition, either of family or community. Some of my classmates' families had lived in the area for generations. They were connected to the land through accumulated ties of knowledge and practice, through family myths and stories and shared experience. Mine were newly minted. But already in the 1960s, shared communal connections to place were fraying. Extended families broke into nuclear families and moved to other villages, or other countries. Sons and daughters of farmers or woodsmen or charcoal burners became electricians or bankers or teachers. That accumulated knowledge and experience no longer had relevance and was no longer transmitted from generation to generation. My personal experiences had to stand in for communal connections, but they were powerful enough to engender my own sense of belonging. The emotional connection to place welled up in that poem and remain with me. I think of myself as a child of that valley with its woods and ponds and views of heathland on distant hills, and to this day I achieve a repose when walking there that I experience nowhere else.

Simon Schama, the art historian, writes that landscapes are 'built up as much from strata of memory as from layers of rock'.[3] These accretions of time and place create the values and perceptions, and the institutions and practices, of communities that are as much part of a place as are its physical geography of mountains, valleys, serpentine rivers and the myriad plants and animals that inhabit them. They are also the texture and grain built up or worn down by soft breezes carrying the scent of pollen, the dust of a harvested field, or the flat light of an autumn afternoon during a single lifetime, less than a lifetime, a childhood. It is these familiarities and the memories of them that led me to the paths I followed through the woods of my home – and which, in time, led me to the trails through the savannahs and forests of Africa. And it was my belated recognition of the centrality of them to my very being that led me to understand that they must be at the heart of our endeavours to protect our natural world.

---

3 *Landscape and Memory*, Schama, S. (1996) London: Fontana Press, p. 651.

1

Setting out to save the world

It was in 1981 that first I saw the rough fabric of hills and valleys that make up the area called Lake Mburo, pinned together at their centre by the lake itself, a jade brooch in a dull gold setting. This subtle, subdued landscape would become my home for ten years, and in time I would come to call it simply 'Mburo'.

We were traversing the southern districts of Uganda, heading west around Lake Victoria and on to the borderlands with Zaire.[4] We had left behind the flaking facades of Kampala's streets with their twisted and rusted road signs and potholed roads, the helter-skelter drives on lamp-less streets and unsettled nights of random gunfire and alarms. Just minutes from the city we were immersed in a landscape so verdant that it seemed like a hallucination.

June marks the start of the long dry season,[5] and we drove west under profoundly blue skies, anvil clouds hanging above the horizon, threading forests and wetlands on a broken road of powdery red dust and sticky black tarmac. We drove through farmlands of fresh green maize and beans in ragged fields, sweet potatoes' blue-green leaves sprawling on mounds of rich red soil, and plantations of waxy-leaved banana trees. Glimpses of Lake Victoria sparkled in the distance.

We passed through cool forest patches, looking up at the raucous calls and heavy wingbeats of black and white hornbills labouring over the tree canopy. We crossed swamps of floating grasses and massed papyrus flower heads that wafted damp, organic smells through our windows. The frame of our frog-eyed Landcruiser vibrated as the wheels

---

4 The country is currently called the Democratic Republic of Congo or DRC. It started life as the private estate of King Leopold II of Belgium, which he named, apparently without any sense of irony, the Congo Free State. It became the Belgium Congo when removed from the king by the Belgian government, when the appalling treatment of its people was revealed. The Congo became an independent and democratic nation in 1960, but in 1965 was seized by Joseph-Désiré Mobutu, who ruled alone as a dictator for over 30 years. As part of his programme to reinstate African culture at the core of the nation he gave himself an African name, Mobutu Sese Seko, and renamed the nation the Republic of Zaire after a local name for the Congo river. When he was finally deposed in 1997 the country became the Democratic Republic of Congo.

5 Southern Uganda has a long and a short dry season, and long and short rains annually.

burbled over the broken road surface, reduced by erosion and years without maintenance to a ragged line straggling between margins of crumbling ochre murram.[6] Crashing through the hard-edged holes in the tarmac sent up jolts like blows that had me hanging onto my seat and bracing myself against the canopy bars.

Quite abruptly we passed out of this landscape of lush fields and banana plantations into a dry land of sparsely covered low hills. Cropped grass stretched away between scratchy bushes on either side of the road, splashed with the orange tufts, like mini mop-heads, of flowering erythrina trees and the dull red hulks of termite mounds topped with scrambling plants. Strolling silently out of the bush, sleek, two-toned impala crossed the road ahead of us. Improbably fat zebras lurched into a canter, plunging alongside us briefly before veering back into the scrubby bush. The occasional topi would run a few tired rocking-horse steps before stopping. The air was prickly with dust and the smell of dung. Flocks of miniature birds, red-billed queleas, fire finches, mannikins, rose from the grasses where they were feeding on seeds with the sound of a wave running up a beach.

The land to the south of the road was occupied by the Lake Mburo Nature Reserve. North lay a ranching scheme developed 20 years before in an effort to create a commercial cattle-raising industry. The fences had fallen, the water pipes were empty, and the metal drinking troughs were rusting into holes. But for the dividing road and the glimpses of infrastructure lost in the vegetation, there was nothing to tell the nature reserve from the ranches; the trotting families of warthogs and the occasional solitary eland certainly made no distinction.

Whatever its designation, this dry land of gentle hills, grassy valleys and granite outcrops that extended from the lakes of Mburo's wetlands in the south to the Katonga River to the north seemed like open terrain, crossed by wildlife, back and forth, the animals migrating in search of grazing and water as the rains came and went. The sight of wildlife roaming freely through the bush land thrilled me. This was what my imaginings as a child had taught me to expect, and my excitement rose as the sights and sounds of the wildlife and the peppery desiccated landscape met the expectations I had held for so long in my mind, my personal dream of Africa.

 [7]

My experiences that day – my first real exposure to a landscape that resonated in harmony with my anticipations – were, I understood later, quite different from those of the people we

---

6 Many of Uganda's roads, especially rural ones, are made of red laterite soils called murram. Though quite stable when compressed, murram roads need to be reshaped regularly. Tarmac laid over a foundation of murram tends to erode from the sides, resulting in an ever-narrowing strip of tarmac down the middle. Drivers approaching head on are forced to play chicken, driving directly at each other before swinging half-off the road to pass.

7 The motif used as a separator here and elsewhere depicts the candelabra tree, *Euphorbia candelabra*, common in the Mburo area. This stylised design is one of the traditional Bahima black and white decorations that represent some of the naturally occurring elements of their landscape.

passed on the road; the farmer pulling a reluctant goat to market; schoolchildren in their bright uniforms running home; the cattle herder watching his animals grazing the spiky pastures. Although for a moment as we crashed along the bumpy road, we occupied the same time and space, seeing the same sights, hearing the same sounds, smelling the same air, our experiences were dissimilar. We might as well have been occupying an entirely different land. This in retrospect is obvious, perhaps, but at the time the significance of this truth for the work I had come to do failed to register with me, let alone occupy my thoughts as it should have.

I had set out from England just a few months before, with the express intention of helping to save Africa's wildlife. But I had barely considered the people I would meet there. I was focused on the world of nature that I would find, anticipating how I would feel to be part of it. Once in Africa, though, setting my feet on the first rungs of a ladder I am still climbing, I realized that I needed to think carefully about how I would engage with the people I was to work with – and, it dawned on me only later, work on behalf of. I had no expectation that the people I would encounter would have the level of knowledge I had. Nor did I expect them to have the same understanding of nature I had. I did not, in fact, expect them to consider nature to any great degree at all. Most of my friends and family at home didn't. Why would it be different in Africa? I anticipated the need to teach people to appreciate the value of nature. Indeed, that was a large part of why I was going there.

That the local people would have their own knowledge of nature and their own unique understanding of their place in it; that they would have relationships with nature and draw on experiences unconnected to my own and likely incomprehensible to me; that their visions would run parallel to my own but be as compelling and meaningful to them as mine were to me – these thoughts I had not considered. It would be years before I began to think about the significance of these differences and consider their importance for my endeavours to protect nature, and even longer before I recognized that they needed to be at the heart of my efforts on nature's behalf.

As we drove through this beautiful land, we began to encounter herds of cattle carrying what looked like enormous ivory tusks on their heads. They were tall and elegant, and they swayed as they walked. I was perhaps even more excited at seeing these wonderful creatures than at the wildlife we encountered; they were so completely unexpected. My books and documentaries had readied me for the sights and sounds of Africa's wild animals, but they had not prepared me for anything like these cattle. No one in Kampala had mentioned that I was sure to see herds of them on my journey, yet suddenly they were all around us. Their chestnut hides and huge glowing crescent horns seemed to reflect the tones and contours of the land, the cattle seemingly the essence of place, drawing their inspiration from it.

Ankole cows moving through wooded grassland in Mburo

'What are those?' I asked, Rob, who was driving. He ignored both me and the cattle, concentrating on the road. He was the regional representative of the Worldwide Fund for Nature – WWF. He was focused on wildlife, not cattle. As I had only recently started work as his assistant I asked no more questions, but this first encounter with these astonishing creatures would turn out to be an important moment for me.

The guardians of the herds, resting under trees or leaning languidly on their sticks, seemed to appropriate the pale drowsy heat of midday – languid, that is, until I raised my camera. Immediately they burst into fierce gestures and angry cries, and the cattle were whooped and waved into the thickets to save them from their images being stolen.[8] We were in the District of Ankole, and it turned out that these were the famous long-horned Ankole cattle, which nobody had thought to mention to me.

To the east and west of Ankole rainfall is high. Those lands had been covered with forest until the trees were swept away by fields and the advancing canopy of bananas, leaving standing just the occasional fig tree or forest giant to tell of what had been. But the lands of Ankole are different, and had never been forest. Rainfall is so low, and evaporation during the long hot days so high, that geographers and botanists making their detailed

---

8 Pastoral people are often reluctant to allow photos to be taken of their cattle, fearing that strangers who over-praise or pay too much attention will endanger them. Herders are also fearful of tax collectors. An exercise was carried out to count the herds of pastoralists living within Lake Mburo National Park, but their reluctance to allow government officials to count their animals was not easily overcome. In Uganda, since the days of Idi Amin, there has also been a fear of strangers taking photos, as this was often a preliminary to someone's sudden disappearance.

observations and records classify the area as semi-arid; not truly arid, not a desert, but marginal all the same and certainly too dry to be sensibly considered farmland.

Ankole is for the most part covered by sparse open woodlands and tangled thickets of short, thorny trees and shrubs rather than the tall, deep tropical forests to the east and west. The rainclouds that form over Lake Victoria, just 50 kilometres away, fail to reach Ankole, while the clouds that mass over the forests of the Congo Basin in the west, forced to rise by the Rwenzori Mountains, drop the last of their rain on their slopes. The result is a band of wooded savannah not more than 100 kilometres wide. It is called 'the cattle corridor' to distinguish it from the well-watered farmlands, and has long been the land of pastoral peoples who traverse it with their herds. The Bahima, part of the Banyankole, the people of Ankole, have occupied the area for 500 years. They call it *Karo Karungi*, the Beautiful Land. It is the land of their ancestors, it is their homeland, and it is the home of their beautiful cattle.

I felt a comfortable sense of luck and privilege to be bumping along the road through the bush in a sand-coloured Landcruiser just months after leaving England – and I was indeed lucky. With no job and no connections, I had travelled to Africa with the idea that Nairobi, the centre of East Africa's safari industry and the efforts to conserve the wildlife spectacle on which it depended, would be the right place for me to start out. Chance and a series of somewhat engineered accidents led me to Rob who, with surprisingly little hesitation considering my lack of experience, took me on as a volunteer on his freewheeling operation to support Uganda's national parks and game reserves. That was luck indeed, and a far cry from the formal process of volunteering and internships that today's young aspirants must navigate.

In 1981 there were just three national parks in Uganda, though there were also many game and forest reserves. All were struggling to recover from eight years of misrule by His Excellency, President for Life, Field Marshal Al Hadji Doctor Idi Amin Dada, Lord of All the Beasts of the Earth and Fishes of the Seas and Conqueror of the British Empire in Africa in General and Uganda in particular; Idi Amin to most.

Prior to Amin, the national parks had been at the heart of a flourishing tourism industry. Murchison Falls National Park, where the full span of the Nile River squeezes through a narrow cleft to crash like a steam engine down to a cauldron of boiling water and basking crocodiles, is an extraordinary place, and was a favourite amongst those early visitors. Queen Elizabeth National Park, on the border with (the then) Zaire, in the middle of a vast landscape of forest, savannah, lakes and mountains, [9] and boasting herds

---

9 The much larger Virunga National Park – it covers 7,800 km² compared to the 1,978 km² of the Queen Elizabeth National Park – in the Democratic Republic of Congo includes the Virunga volcanoes in the south and the Rwenzori mountains in the north. Established in 1925, it was originally named after King Albert I of Belgium.

of elephant, buffalo and hippo, included the vast Lake Edward and glimpses of snow-capped mountains. Kidepo Valley National Park, meanwhile, was isolated and hard to reach, set on the northern border with Sudan.

Together these parks attracted more than 160,000 visitors a year.[10] Within months of Amin taking power, as reports of the killings filtered out, Uganda was taken firmly off the tourist trail. Though tourists continued to flock to neighbouring Kenya and Tanzania, visitor numbers to Uganda fell. Amin's decision to close the country to international tourists two years later, arguing that the parks should be for Ugandans, not foreigners, was nothing more than posturing; the tourism industry, always sensitive to the least smack of uncertainty or unpleasantness, had abandoned Uganda.[11]

After that, the government funding for the parks quickly dried up, and the international conservation bodies went elsewhere too. As the madness of Amin's regime spread more widely, rangers were unable to protect the parks from local hunters, let alone from warlords and soldiers hunting for fun and profit.[12]

And then, although the Tanzanian invasion that deposed Amin in 1979 returned Uganda to the rule of law, for a while at least, the plight of the parks actually grew worse. To recoup the costs of its invasion and occupation, the Tanzanian army set about hunting and trading wildlife meat on an industrial scale. Buffalos, hippos and the larger animals all but vanished. Elephants had already disappeared, shot for their ivory or having fled across the borders into Zaire and Sudan.

Back in the 1960s researchers working in Queen Elizabeth National Park had found that the Mweya Peninsula, a diamond-shaped isthmus projecting into Lake Edward, carried the highest concentrations of animal biomass – wildlife – ever recorded.[13] Herds of buffalo and Uganda kob grazed the peninsula, hundreds of hippos spent their days

---

10 Murchison Falls Game Reserve, established in 1926, was the backdrop for *The African Queen*, the film made in 1951. Humphrey Bogart, who won an Oscar for playing the steamboat's irascible and drunken captain, and the rest of the cast and crew, stayed for weeks in the reserve. This was unusual, most films being shot on sets rather than on location. That the reserve could host the stars of Hollywood indicates how well developed its tourism infrastructure was; Bogart and his co-stars were not accustomed to slumming it. Perhaps part of the drive to film on location was the obsession of the film's director, John Huston, with shooting an elephant. The tragic consequences that resulted were the subject of a 1990 film, *White Hunter, Black Heart*, starring Clint Eastwood as the increasingly obsessive Huston.

11 Tourism is often seen as the obvious solution to providing incentives for conservation, but it is a fickle industry. Tourism can provide economic arguments for governments and communities to support protected areas, but the implications are grave when tourism fails, and there are many reasons why tourists might stop coming, most of them outside the control of the industry or the country.

12 Alfred Labongo was the Warden Law Enforcement at Queen Elizabeth National Park in 1981 when I arrived there. He described how soldiers had come for him during his time in Murchison Falls. They took him at gunpoint to the bridge at Karuma under which the Nile rushes over a series of rapids with mesmerizing power and speed. Rather than take the inevitable bullet he threw himself from the bridge, miraculously surviving the fall, the rapids and the crocodiles.

13 Biomass, simply the quantity of organic matter, is a measure commonly used to describe the ecology of a location and is applied to both plants and animals. A classic piece of research on the ecology of Tsavo National Park in Kenya, for example, demonstrated a direct relationship between rainfall and the quantity of animal biomass. *The distribution and abundance of the large herbivore communities of Tsavo National Park (East), Kenya*, Cobb, S. (1976) PhD thesis, Oxford University, UK.

basking around its shoreline, emerging at night to crop the grass into fine hippo-lawns, and elephants wandered down the narrow neck of land to browse on trees and shrubs. In fact, the pressure on the park's vegetation was so intense that ecologists recommended one of Africa's first culling schemes.[14]

By the 1980s did anything of this famous spectacle remain? The comparative stability of the first years under Milton Obote, who replaced Amin, encouraged international conservation organizations to return to see what had happened in the parks and reserves during the years of chaos. A survey was organized to help conservation professionals, Ugandan and international, decide what could be done.

In the years following the independence of Africa's many nations,[15] scientists and conservationists watched with dismay as their plans for the development of tourism, hunting, cropping and even the domestication of animals such as eland and oryx, foundered. East Africa, where the greatest wildlife spectacle on earth still persisted, was the focus of particular attention. In the run-up to independence, many American and European organizations were anxious to build cadres of committed and qualified wildlife professionals and build local support for conservation. To achieve this they set up programmes of study, research and public education.[16]

In Uganda the possibility of continuing this work, which had faltered under Amin, generated interest in rebuilding its parks, and it was this post-Amin flurry of activity that provided my first job in conservation. My entrée to the world of protected areas came when I joined Rob's project to survey Uganda's game and forest reserves. Our base was Queen Elizabeth National Park, which lies within the great Western Rift Valley.[17] The survey would investigate the state of the reserves in western Uganda. The job also delivered my first sight of Mburo, from the ground and the air, as well as my earliest inklings of the

---

14 This incredible concentration of herbivores was an outstanding attraction for tourists, but a worry to the ecologists of the Nuffield Centre for Ecology based in the park. Fearful for the park's vegetation, they devised Africa's first culling programme to reduce the number of hippos and elephants. It was a controversial decision to shoot animals in a national park, but thousands were killed in the name of conservation. Emerging differences between *laissez faire* approaches to management in East Africa compared to interventionist approaches in southern Africa led the call to cull elephants in Kenya's Tsavo National Park in the 1970s to be rejected, and to the reported death by starvation of over 3,000 elephants.

15 The process by which African states were made independent in the 20th century was almost as rapid as the process by which they had been colonized by European states in the 19th century. No less than 32 African countries, 60 per cent of those that exist today, became independent over the ten years of the 1960s.

16 The African Wildlife Foundation, formed in 1961, the year before Uganda gained independence, was originally named the African Wildlife Leadership Foundation, and concentrated on building a generation of trained conservation professionals and on educating the African public on the importance of nature conservation. Both these endeavours entailed purveying a cultural perspective of nature, what it was and why it was important, that went far in creating a conceptual divide between African conservationists and the African people on behalf of whom they were working.

17 The park was named to commemorate the visit of Queen Elizabeth to Uganda in 1954. To this day the park sports signs directing you to the Royal Circuit, the Royal Mile and the Royal Pavilion. When I first lived there the park was named the Rwenzori National Park after the nearby mountain range. The government of Milton Obote revoked every act of Amin, even sensible ones, and so the park became again, and remains to this day, Queen Elizabeth II National Park.

tensions between a western-based vision for nature conservation and the consequences of pursuing that vision.

From the grass airstrip that cut across the middle of the Mweya Peninsula our pilot[18] would broadcast his intention to take to the air, just in case any airliner passing high above might be paying attention. Then he would fling us up and out across the turquoise waters of Lake Edward before wheeling north or south towards one or other of the reserves we were to survey.

We flew over the game reserves and forest reserves – the national parks had already been covered – counting herds of cattle, herds of sheep and goats (lumped together on our data sheets as shoats[19]), huts and fields and, of course, any wildlife we saw. It was my job to record what I saw through the window on my hand-held tape recorder – the first time I'd held a Sony Walkman in my hot hands[20] – calling 'shoats shoats 20', 'cattle cattle 6', 'huts

Left: The Aerial survey team, Dr Keith Eltringham, Dr Rob Malpas, Justus Tindygarukayo and the author

Right: Putting the survey plane safely away, with help from our friends

---

18 Our pilot, Dr Keith Eltringham, a professor at Oxford University, had been a long-serving ecologist at the Nuffield Unit of Ecology, which after independence became the Uganda Institute of Ecology. Like a number of his profession, he was a first-class aerial survey pilot.

19 Sheep and goats are very similar in appearance but are classed as separate species in separate genera and do not produce fertile offspring on the rare occasions they interbreed. Goats are browsers, preferring leave and twigs, while sheep are short grass grazers. That aside there is little to distinguish them, certainly from 300 feet. The noted biologist, George Schaller, scaled the Tibetan plateau to decide whether the blue sheep or Bharal, was actually a goat by observing its mating behaviour. He concluded that it was a goat with sheep-like traits, which I am sure settled the argument completely.

20 The iconic Sony Walkman had appeared in shops the year before. That we were using one rather than pencil and paper demonstrates how quickly researchers take up new tools, converting them as needed to their work. Wildlife researchers have been equally quick to find uses for satellites and drones. When not using the Walkman for research we used it to listen to music, of course, and I cannot listen to London Calling by the Clash or Beethoven's 4th Piano Concerto without being transported to the back seat of that small plane.

huts', and less frequently, but with an excitement I could not keep out of my voice, 'buffalo buffalo!' or 'elephant elephant!'.[21]

The initial glamour of these flights did not last long. The method of counting was gruelling. I was not looking casually out of the back window and counting the moving shapes fading away in the hazy distance or disappearing under the plane, or glimpsed between the wing struts. This was no safari. It was a survey governed by strict methods that required us to count what we observed while flying over carefully calibrated strips of the ground below. Pairs of lightweight fishing rods were fixed under each wing to make parallel markers, while on the plane's back windows we drew more pairs of parallel lines. When the paired markers were lined up and when the plane was flying at exactly 300 feet above the ground, we could calculate the area running between the lines as we flew. This was our transect, and when all our transects were added together they made up the sample of the reserve we had counted over. To estimate the total number of animals or huts in a reserve we divided the numbers of whatever we had counted by the total area of the strips we had counted over – our sample – and multiplied this by the area of the whole reserve.

It was the pilot's job to keep the plane at 300 feet, his eyes fixed to the radar altimeter to maintain our height as the land rose and fell below us and fly a straight line for a measured distance. It was our job to keep our eyes fixed on the ground below and count. We would zigzag back and forth across the reserves, our heads fixed in position and our eyes scanning the strictly delineated strip below, until we had surveyed a big enough part, ideally 10 per cent. This was a sample count, rather than an attempt to count everything. It took an effort of will to ignore the herd of eland, or the single reed buck – perhaps the only wildlife spotted all day – as they drifted past just outside the limits of each transect.

By the end of the day I would have a crick in my neck, eyes sore and watering from the glare, and a headache from the hot, over-breathed air of our Perspex bubble. At least I had a solid stomach, unlike my colleague, Justus, a quiet, thoughtful man from the Game Department with a short curly beard and large eyes, who would be sick as often as not before we reached home.

'Sorry,' I would say gently.

'Sorry,' he would reply sadly.

And the sickly smell in the confined space rose up, challenging the rest of us to hold onto our breakfast.

Though it was exhausting work, especially for the pilot, I found it exhilarating to be high over the crenellated lines and folds of the Western Rift Valley. One moment I would be peering down into scattered volcanic craters, some with forest at the bottom, some with lakes, others covered by smooth expanses of grass; the next we would be crossing wind-

---

21 We called out the word twice before the number to make sure that as we pressed the record button we didn't accidentally miss out or half record the key data, the quantity we had counted. The technology was new, and we didn't want to make a simple mistake that would require a survey to be re-flown, wasting fuel and time, not to mention personal resolve; every morning it got harder and harder to climb into the back seat of that plane.

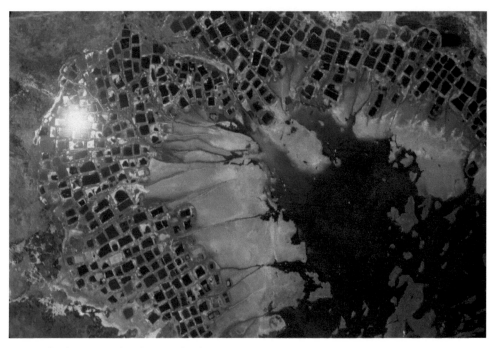

Mineral salt artisanal industry of Lake Katwe, QENP

scuffed lakes streaked with lines of foam and dotted with canoes or skirting the ridges of mountains with rivers rushing down to the dusty planes. Day after day we set off to swish backward and forward across these beautiful lands. Goats would run from the noise of our engine, and children would rush from thatched huts into bare, swept compounds, and jump and wave and shout silently up to us.

The forest reserves were remarkably intact. Lines of exotic trees, eucalyptus or pine, demarcated many of their boundaries, and though farms had advanced up to these, there they had generally stopped. Flying the forest reserves was easy work, as there was little to count; no animals, domestic or wild, could be seen under the dense forest canopy. Occasionally we flew over gaps, small pit-sawing operations hidden deep in the forest, or the rough shelter of a doughty hunter just visible in a tiny clearing far from any village. But the game reserves were different. Though the huts and fields also mainly stopped at their boundaries, unmarked though these were, we were kept busy calling 'shoats shoats' and 'cattle cattle' as the grazing herds covered the lands below us. There was, though, a depressing lack of wildlife, and we would often return home having seen hardly a wild animal all day.

One bright morning we flew east to Lake Mburo Game Reserve. Even before we reached the reserve it was clear that this count was going to be different. We began seeing animals; cattle certainly, herds of the brown beasts carrying their sweeping white horns,

Left: Western hills of Mburo hills from the air during the 1981/82 aerial survey
Right: Reeds, papyrus and sedges form floating mats of wetland vegetation

but wildlife as well. Zebra and impala scattered beneath us, groups of topi rocked through the acacia trees and bushbuck stood frozen atop anthills. For the first time during the surveys, I was experiencing something approaching the wildlife spectacle I had been conditioned to expect.

But Mburo was not like the parks of the television documentaries I had grown up watching, full of prolific and burgeoning nature but without any sign of human life. Instead, here, scattered throughout the low hills and grassy valleys, were patches of bare ground linked by fine lines like the radials of spiders' webs. These were homesteads, the *eka*[22] of Bahima pastoralists, the cattle keepers of Ankole, and it was their cattle that had trampled these patches bare of grass. Huts, and enclosures for calves of different ages were arranged around their edges, sometimes contained by thick barriers of acacia bush, cut and piled.

The paths showed the daily movements of the cattle as their herders walked them to graze in the morning and back to the eka in the evening, where they would stand or lie quietly through the cool nights; the paths also showed the patterns of people visiting their neighbours. The homes of the Bahima were a lot easier to see from the air, I would discover, than from the ground; at ground level they are almost invisible until you stumble into them, so well do the branch, grass and mud structures merge with the woodlands in which they cluster and from which they have been built.

We flew our transects over Mburo from the hard black line of the Masaka–Mbarara road, ruled east to west across the tawny grey bush, to the green, gold and yellow mats of floating swamp vegetation, looking like bacteria spreading on dishes of bitter chocolate, that marked the southern edge of the reserve. Looking down on this exotic landscape

---

22 The Afrikaans word for a fenced enclosure for livestock and people, *kraal*, has become the generic for the households of pastoralists, which inevitably have many common features. The Maasai word *manyata* is almost as well known. The Runyankole word, *eka*, comprises the shelters of people and calves and the grounds on which the cattle gather at night, all traditionally surrounded by a fence or barrier against human and feline predators.

I felt a surprising resonance with what I was seeing, a sense of familiarity, as though I recognized and understood it.

This did not seem like a landscape of primeval emptiness, the exclusive domain of wild animals. It was certainly not the mythical wilderness that as a child I had longed to be immersed in and become part of. It seemed, nonetheless, to call me. I felt an unexpected connection, as though I could have a place there alongside the pastoralists whose home it so clearly was. Like the beautiful lands I had grown up in, the countryside of southern England, the heathlands of Ashdown Forest, this was also a place of occupation, a place with connections, a place with stories. It was a homeland, a land of its people, a land created in form and function by those that lived in it, worked it, and walked through it. As I looked down, I understood that this land was an embodiment of the history of its people, a place with owners, a place I have since learned to describe as a cultural landscape.

Parachuted into the midst of a rearguard action to save what was left of Uganda's wildlife, I found myself living a life so close to my childhood dreams that I frequently had to remind myself why I was there. I was not there for my own satisfaction but to help deal with the complicated, conflicted and challenging reality of wildlife loss. Despite the problems the parks and reserves had experienced under Amin and during the occupation of Tanzanian troops, Uganda's infrastructure for conservation had remained largely intact. The system of parks, wildlife sanctuaries and game and forest reserves was still in place, even if many of the animals and trees had gone. The institutions that managed these areas had emerged from their decade-long struggle still firmly wedded to the principles of conservation. That meant, then, at least establishing protected areas and making sure no one violated them, resulting in a separation between the local people and nature.

The rangers, guards and officers of Uganda National Parks and its sister organization, the Game Department, policed the protected areas with their rifles and uniforms, marching and saluting. Their organizational culture perpetuated and reinforced a perception of competing interests, painting a picture in which they were cast as the righteous protectors of nature. Their neighbours, meanwhile, carried all the characteristics of the shifty, swivel-eyed poacher I had absorbed as a child from history and myth. Wardens and rangers were at war with the uneducated, unsupportive and uncompromising public, and even with their leaders.

The forces of conservation might have been forced into retreat for a decade under Amin, but now the relief column had arrived, headed by the international conservation charities. Now the protectors of nature were ready to reassert themselves and push back against those encroaching on their moral authority as well as those who encroached on the physical integrity of the parks and reserves.

During the 1970s the protection of the parks had been seriously compromised by the withdrawal of government funding, so the wardens and rangers responded to the erosion of their authority with reluctant pragmatism, softening the way they managed the parks. All sorts of informal agreements were entered into with local leaders and warlords to help ensure that the parks and their officers survived.

In effect, access to the resources of the parks by a range of parties, from local people and politicians to warlords that ruled many areas, was bartered for a level of tacit acceptance of at least some of the park managers' rights and responsibilities. This can be dismissed as corruption, and individual wardens and rangers certainly did benefit directly.

But it might be fairer to regard it as a practical response to a situation that required the national park ideal to be compromised in order to allow the parks to survive. Relinquishing the exclusionary nature of the traditional national park helped the wardens retain a level of control over them, preventing what might otherwise have been the wholesale exploitation of their resources and, ultimately, their destruction.[23] It provided income to cover management costs, including feeding staff, and helped keep the parks' boundaries intact. The cost was that some of the wildlife had to be sacrificed and some conservation principles, relaxed. The wildlife, grazing and other resources such as firewood and thatch were traded by the wardens to retain control of the land– which, as they knew, once lost is lost forever. The deal also implied acceptance of the parks as legitimate and distinct from the rest of the land, and of the wardens' authority over them.

Perhaps, more controversially – and I perceive that this was not intended by the wardens – the actions taken by them can be thought of as steps towards a partnership with local communities. Even if partnerships were based on the assumption on one side and tacit acceptance on the other that the parks belonged to government, they were partnerships nonetheless. This was because the wardens and community leaders had to sit down together to discuss the challenges they were facing, reach agreements that were helpful to both, and make decisions together on how best to take forward what had been agreed. By some ways of thinking, these were the very processes that community conservation projects would seek to establish 20 years later, and which have been a significant part of the management of parks and reserves ever since, significant at least in theory.

This spirit of partnership, if that is what it was, did not persist. With the return of the rule of law, or at least the return of a government presence, the wardens reasserted their sole control over the parks. With international and government support they once again had the means to exert their authority and reaffirm the policies and practices of protection and exclusion. They had funds to buy fuel for their vehicles, to purchase rations for their patrols, and to load their guns with bullets. The steps towards collaboration, taken unwillingly perhaps, forced by circumstance rather than choice, were quickly reversed.

---

23 Many of Uganda's forest reserves were lost during this time as a result of the breakdown of law and order that allowed them to be encroached and converted into farmland. Indeed, Amin's 'double production' policy actually encouraged communities to convert forest reserves to farmland in an effort to reduce poverty.

Exclusionary practices were put in place again, confirming the dominance of the park managers – and re-igniting the conflicts with their neighbours.

It was not concern over the idea of separating people and nature, a rather tenuous concept at best, that first began to worry me; it was the unavoidable implications of this separation for relations between us, the conservers, and the people for whom we allegedly conserved. My first months in Uganda and Queen Elizabeth National Park were so true to my childhood phantasy I could scarcely believe it was real. I lived in one of Africa's greatest national parks. I spent my days amongst lakes so large that their further shorelines were invisible and generated waves like the ocean. I lived under snow-capped mountains, and in forests and savannahs full of Africa's most totemic wildlife.

I grew accustomed, though never insensible, to the drama of dropping down the precipitous Banyaruguru Escarpment on the switchback road, crossing the Kazinga Channel where swifts and swallows screamed across the wind-scuffed water, then turning off the main road at Katunguru, and heading up the long rough track to the park headquarters at Mweya, exotic names that still retain their enchantment over me. Perhaps I would encounter a lone elephant crossing the plains or stop to watch the herons and egrets standing statue-like in the reeds. If rain had cleared the dust-laden air I could raise my eyes to the glaciers of the Rwenzori Mountains flashing in the sun 50 kilometres off and 3,000 metres above[24]

Back in Mweya I might watch lions lounging in the grass as I sat on our veranda at sunset, listening to the gargle and roar of hippos in the shallows below the house. A slight rasping would alert me to the monitor lizard that hunted bats in our attic warily descending the outside wall.

I did not take these things for granted; I marvelled in them and was thankful I was there to experience them. But what I *did* take for granted, to begin with at least, was the entire system of thought, practice and action that protected these creatures in their habitats by keeping everyone out – everyone that is except for the likes of me, who had secured a privileged place in their midst.

These concerns began to demand my attention, insisting that just when everything seemed so right and perfect for me, something was wrong. The first clouds began to dim the brightness of my glorious, dreamlike days. As I drove through the fishing villages – Kasenyi, Kazinga, Hamukungu – that existed as enclaves within the park, all communities that had existed long before its creation, or stopped to buy pineapples and mangos from

---

24 The dry season can be so dusty, and the air so filled with soot from burning fields and bush, that visibility can be terrible. Henry Morton Stanley famously walked right past the Rwenzori Mountains without noticing they were there during one of his expeditions, leaving them to be discovered by his second in command, William Grant Stairs.

Left: Thundercloud over the Kazinga Channel, Queen Elizabeth National Park
Right: Fishing while swifts and swallows hawk for insects on the
Kazinga Channel in Queen Elizabeth National Park

rickety stalls set on road verges outside the park, where people struggled to grow maize and beans for their families, I saw few of the waves and smiles and shouts of '*Muzungu, muzungu*'[25] from the children that I was accustomed to being greeted with in Kampala or other villages. Men working on their tangled heaps of fishing nets, fixing floats carved from *ambatch*,[26] would look up with guarded, grim faces. Women, laughing as they pounded cassava, would fall silent on seeing us.

Driving through the bush in a battered open-backed Land Rover to deliver food rations to outposts, rangers in uniforms, albeit little more than rags, carrying ancient wooden-stocked Enfield rifles, hanging on the back, filled me with schoolboy excitement. Yet I could not fail to see that the people we passed, all too often left in a thick cloud of red dust, saw things differently.

I believed, and believe still, that to protect the things we value rules are needed, and that in order to have any meaning they must be enforced. But when I observed how this played out for the communities living around Queen Elizabeth National Park, I felt increasingly uncomfortable. The rules were not their rules, and they had had no say in making them or

---

25 *Muzungu* is a KiSwahili word applied to foreigners in general but is mainly understood to mean a 'white person'. It literally means 'those that go around and around', or 'those that move around all over the place', an implied criticism of the ill-considered traipsing around of white people in East Africa. These travelled far and fast by local standards, one moment being reported in Zanzibar, the next in Buganda, the next in Dodoma.

26 Ambatch or balsa (*Aeschynomene elaphroxylon*) is a small tree with delicate yellow flowers that is common along the shores of Uganda's lakes and rivers. Its pithy wood is cut into pieces and tied to fishing nets to keep them afloat. After months of use they erode into rounded, onyx-coloured baubles that are a lot prettier than the plastic bottles or chunks of polystyrene that have largely replaced them today.

in how they were administered. It was clear that they had little feeling for or belief in what the park delivered. They were obliged to follow the rules, even though it cost them dear, or face punishment. The whole set of practices and paraphernalia that related to law enforcement – the guns, the orders and commands, the parade ground discipline – made me feel uncomfortable.

It might all be necessary in the context of that specific world of paramilitary policing, but I couldn't believe that in the long run the approach would work. Sadly, there can be no argument against the need for well-led and equipped ranger forces to contend with the massacre and theft of rhinos and elephants in Africa, tigers in Asia, indeed any species anywhere unfortunate enough to carry horns, tusks, hides, scales or bones that fetch unimaginable prices. As long as there are people whose desire for profit trumps any concern for the animals that produce what they trade, or for the sustainability of that trade, or for the lives of people who depend on those animals, they must be opposed.

Without protection, even with protection, as we have seen, the steady and saddening decline of wonderful creatures at the hands of poachers, traders and consumers is almost guaranteed. But rangers do not spend most of their time protecting wildlife with high commercial value from such criminal networks. Most of the policing, perhaps because it is easier and safer, focuses on local people attempting to meet their need for firewood or their wish for bushmeat. The policing mainly affects community members whose interests are very different from those of the professional poacher.

Surely these local demands could, for the most part, be accommodated through partnerships between park managers and local communities. They had been accommodated, after all, just a few years before during the bad days of Amin when a different set of rules, designed to achieve a different kind of protection, had evolved and briefly flourished.

I had grown up wanting to conserve nature because of its beauty and complexity, for its astonishing and endlessly surprising and inspiring invention, and because it made me feel good. But now, looking back, it seems obvious that it was for my own sake, rather than nature's sake, that I believed in conserving it. Even back then, though, I had begun to realize that if conservation was in part at least for me, then it could equally be, and perhaps *should* equally be, for the sake of others too.

Even as this conclusion insinuated itself into my thinking, the idea that the connections between people and nature, whether mine or someone else's, could be the reason for conservation as well as the means to achieve it had not formed. The day-to-day activities of park management that separated the people from the natural world were so embedded, and the outcomes for me so seductive, that I accepted them as normal and inevitable, if not entirely desirable, without thinking about it.

All seemed well until I had to confront my emerging concerns about the consequences of this separation. Why were our neighbours looking at us with such suppressed anger and resentment? Why were they apparently not as enamoured by the thought of protecting nature as I was? Did they not value the natural world that was all around them, that was their heritage so much more than it was mine?

These naïve questions about the people and their values grew in response to the casual language of dominance that my colleagues and I used to refer to the people on whose behalf we were protecting nature. The way we used 'they' in 'they don't care', or in 'they don't understand' provided the footing for concluding that 'they' needed to be educated, advised, kept well out of the away and, if necessary, opposed. We were in no doubt that these people, our neighbours, our proposed partners, the very targets and purported beneficiaries of our work, had to be managed and controlled. The idea that they could be engaged with, listened to, learned from, was not part of conservation practice, of what we thought and did.

My growing doubts and discomfort forced on me the conclusion that conservation could not succeed in an atmosphere of mutual suspicion and antipathy. There had to be a partnership between the parks and the people living around them. I was not, of course, alone in thinking this. Nor was it a new idea. The negative implications of the exclusionary model of conservation were increasingly clear in post-colonial Africa and elsewhere. Communities were challenging protected areas, politicians were questioning their value, and campaigners were raising concerns over infringements to human rights in the name of conservation. Academics and conservation practitioners were also beginning to look for different ways to deliver conservation.

I did not at this point question the importance of pursuing the conservation of nature as an undertaking – that would come later – but I increasingly wondered about the way we were attempting to do it. I became convinced that we needed to change our ways to have hope of success.

In the early 1980s raising questions about the nature of parks and the practices used to protect them could be difficult; everyone suddenly seemed to support the idea and urgency of saving wildlife, and many seemed to have formulated a clear if simplistic view of the issue. On my visits home, I was a veritable hero, praised for my commitment. Considering that half my age mates in Thatcher's Britain seemed to have become bankers or commodity brokers, by comparison I seemed to embody selfless integrity.

Conservation was lauded unhesitatingly as a complete and unquestionable good, and those that devoted themselves to it must be honourable and self-sacrificing. But after just a short time in Africa, I could no longer see conservation in such exclusively positive terms, or consider my own involvement in it as selfless. I was already beginning to see my world in uncomfortably numerous shades of grey rather than the blacks and whites I had seen just a short while before.

Looking back to my time as a student in the 1970s, when I had first learned to think about nature in terms of science, and then to those early experiences in Uganda, when I was forced to think about conservation in practical rather than nebulous and strongly personal terms, I fear that I accepted too easily ways of thinking that were at odds with my feelings. I took them up and followed them with energy, but in retrospect they were contrary to my outlook and did not sit comfortably with me.

This happened in part, I think, because I didn't habitually ponder nature, let alone my relationship with it. The pathways that my thinking followed, and therefore my understanding of conservation, were opened by my schooling, by my studies, and by the beautiful theory of evolution as I encountered it. My first professional experiences widened these paths. Even if, despite my best efforts, I failed to become a scientist, that way of thinking and forming knowledge influenced how I experienced nature and determined how I began to think about and pursue its conservation.

Under the expectations of teachers and employers I allowed the sense of my personal connection to the natural world to slowly erode. Deposits of acquired knowledge, practical experience and accretions of solidarity with the embattled champions of nature slowly but steadily obscured it. This was not just a personal loss, though. My increasingly formalized, information-based understanding of nature allowed, encouraged, perhaps even required me to overlook the myriad relations that the people I worked with in countless villages around protected areas had with their own natural world.

By forgetting my feelings for nature, I came to assume that others too had none. The demands of conservation as we practised it required me to bring these people to the same state of knowledge and scientific understanding that dominated the conservation endeavour and my own thinking. My feelings became buried so deeply that I hardly remembered them, and never talked about them, certainly not in the company of my colleagues. I had replaced feelings with the material, rational and dispassionate construction of nature that was the essence, the very principle, of modern conservation.

As my career developed and I moved from Uganda to work in other parts of Africa and then to Asia, the more time I spent in the most beautiful places in the world, the more of nature's flamboyant glories I enjoyed, and the more privileged I felt to be having these experiences, the stronger my conviction grew that they had to be safeguarded. I believed that the future would blame us for every creature or plant, great or small, that we lost through our lack of care and attention, or through our greed. But the longer I worked in my small way to help save some of them the more concerned I became with the methods I used, and the more evident it became that I was failing and must fail entirely.

Most of the time I had my head down, though, struggling with the challenges of a particular site or species or project, sometimes having small successes, but often achieving only small failures. But when considered squarely, the bigger picture made it impossible to ignore the global scale of the problem. All the graphs showed species, populations, habitats, going down. Surely that meant we had to question the direction we had taken? I was not ready to abandon the idea of protecting nature and join those who behaved as though they didn't care about it. Nor could I align with those that cared so much that they could justify any steps to protect it. But that we needed to be doing things differently seemed clear. More of the same was no longer an option.

2

The rise and fall of a national park

The park and the people

Lake Mburo National Park is an emerald glowing in the subdued tawny expanses of the dry lands that sweep up from northern Tanzania and on through to Karamoja on Uganda's eastern border. Swamps and lush valleys burst on eyes accustomed to searching for more subtle signs of life in the desiccated woodlands. Seen from the air, the inundated heart of the park appears as a profusion of organic colours and shapes, mats of vegetation forming discs of intense colour, green and gold, floating on the tannin black of open water. Compared, perhaps, to the great mountains and lakes of the Western Rift Valley, or the unbounded savannahs of the Serengeti plains in the south, Mburo is a more intimate landscape, built on a smaller, more comfortable scale. It is a place of tree-fringed lakes, some no more than pools, set within modest hills and valleys shaped by soft swells and curves.

The Rwizi River fills a broad, irregular basin to create a permanent wetland of papyrus, cypress, sedge, reed and water lily before finally seeping into Lake Victoria. This extensive swamp and the 15 lakes it embraces form the southern boundary of the park and are the linchpin of Mburo, its life force. As the sole permanent water source in the area, they form the focus of migrations of wildlife as they follow the pastures that ebb and flow with the seasonal rains. Historically the Bahima pastoralists followed this same pattern when their herds required it. Tall, closed canopy forest grows close to the river and wetlands in places, dense, closed thorny thickets grow on steep dry hillside, and mixed open woodlands cover the gentle hills punctuated by occasional rocky tors, while grasslands dominate the open valleys.

Across these varied habitats live numerous animals and birds. The park claims over 68 species of mammal. The large charismatic animals are the most evident. Herds of buffalo, zebra, impala are seen everywhere in the park during the dry season. Eland,

Lake Mburo viewed from the western Kigarama hills.

Swamp growing papyrus flower head.

Africa's largest antelope, with tawny flanks and corkscrew horns, are less easy to find but occasionally congregate to form herds 200 strong. Topi and bushbuck are more evident, standing as they often do on termite hills, but the smaller reed buck, bush duiker and oribi can be easily overlooked. As far as the smaller mammals are concerned, Mburo is remarkable in having six species of mongoose. Leopards abound in the park but, they like the other predators, are rarely seen. Mburo's 300 bird species cover the range of bushland, woodland and wetland species, including many that are rare or threatened.

The eastern part of the park curls around Lake Mburo. Low hills are strung together like knuckled fingers of a hand, separated by valleys that run with slow floodwaters during the rains. They toss with the golden

21

Left: Eland in Mburo, generally seen in small groups, Eland herds of 200 have been seen at Mburo. (copyright Connie Bransilver)

Upper right: Crown cranes, the national bird of Uganda, in grassy valley bottom in Mburo

Right: Two male buffaloes in the wooded hills above Lake Mburo

tassels of grasses, *themeda, loudetia, eragrostis*, which spring up as the rain sinks into the peaty soils, to be grazed down to a spiky sward or burned off into blackened clumps as the dry seasons advance and the heat intensifies. A broad thumb of land separates Lake Mburo from a range of hills that rise abruptly in the west. These mark the start of a series of rocky ridges and broad valleys that cross the western extension of the park and run down to the floodplain of the Rwizi River, whose brown waters flow amongst a forest of fever trees.[27]

Mburo lies at the heart of the Beautiful Land. Though the boundaries of Nkore, the kingdom of the Ankole people, changed many times, the Mburo area, known to the Bahima pastoralists as Nshara, lay at its geographical and emotional core. Despite my

---

27 Science can be exasperating. Its exactitude takes no account of feelings. The fever tree, a name lodged in my childhood memory, was probably the first tree I learned to identify in Africa, its unmistakable lime-green bark giving it its name, *xanthophloea*, of the genus *Acacia*. It was called the fever tree because it was thought to cause fever. This belief arose because it grows in the marshy areas that are also habitats for malaria carrying mosquitos. Later I learned to recognize the whistling acacia that sports thorns sprouting from enlarged hollow nodules into which ants bore holes that whistle musically in the wind. Flat-top and umbrella acacias were also easy to recognize, and they came to define the African bush for me. The fever tree was one of several acacias in Lake Mburo. Others included *Acacia hockii*, whose yellow pompom flowers fill the air with scent, *Acacia gerrardi*, which looks ancient, gnarled and lichen-covered, even when young, and *Acacia sieberiana*, which carries long curved white spines. Acacia trees were part of my Africa until in 2015 the XVII International Botanical Congress decided that Australia would have the privilege of retaining the name 'Acacia' for its 900-odd species, while Africa was stripped of it. My beloved acacias suddenly became *Vachellia*. Very upsetting.

immediate attraction to Mburo when I first saw it in 1981 and then again in 1988, Bahima perceptions of the area and its worth were very different. Where I saw beauty in the lines of impala walking to water in the evening, Bahima saw beauty in their long-horned cattle, knee-deep in the pastures. Where I treasured the shape of hills falling like a shoulder to the lake, the Bahima cherished the open valleys for their rich grazing and abundant pools. Their emotions, their love for the Beautiful Land, might seem absorbed by the practical matters of cattle-keeping, bereft of a deeper sensibility, but for the Bahima there is nothing more profound than the connection between their cattle and their homeland, and nothing more beautiful than the cattle on the land. A recitation, a form of improvised Bahima poetry, expresses this sense of connection deftly.

> The grass of Rwanyakizhu is burnt,
> Bawatimba is bare,
> The fire has gone to the shores of Lake Kakyera,
> and burnt Ruragara.
> And the cows stand still at Rukukuru,
> As the ripened fruit of Enyonza[28]

Lake Mburo National Park's 260 square kilometres is diverse. There are broken rocky slopes, and granite outcrops emerging from the sides of hills, parched stubbly flats lying over light rocky soils, and lush grasslands growing on rich black soils. There are forest patches in sheltered valleys and impenetrable thickets of thorn and bush. Accounts of Mburo and the larger Akagera ecosystem that exists in nearby Rwanda explain their surprising diversity of plants and animals and the many habitats crammed into relatively small areas – small, that is, in the context of the vastness of the African continent.[29]

These ecosystems, represented by scientists with just a handful of simple terms – woodland, grassland, valley grassland, thicket – are complex. Each one, however, is distinct from the other.[30] In each of them are gathered communities of plants and animals attracted by the particular conditions they provide as a result of the differing geologies and soils that underlie them, the steepness and orientation of their slopes, and importantly, though often discounted, their history of use – or, as will be described later, abuse – by humans.

---

28 This fragment of a Bahima recitation conveys to me a deep sense of beauty and repose in the landscape, and connects the cattle to the place and its changing moods and seasons. Through the allusion to 'enyonza', the poem establishes an historical link to the time of the rinderpest epidemic. During this time, fruit of the enyonza tree were eaten to stave off starvation, and used instead of cattle to pay bride price.
From: *The Heroic Recitations of the Bahima of Ankole*, Morris, H.F. (1964) Oxford: Clarendon Press.

29 Continental Africa is big. A wonderfully effective graphic shows just how big Africa is by placing China, India, the USA, Japan and most of Europe comfortably within Africa's 30 million square kilometres https://blogs.scientificamerican.com/observations/africa-is-way-bigger-than-you-think/

30 Landale Brown carried out assessments of vegetation types throughout Uganda in the 1950s and 1960s. Despite the vast diversity of the species he identified, he represented the vegetation of the entire country by describing just 20 habitat types, each labelled with a simple two- or three-word descriptor.

Left: Hills of the eastern sector of Mburo, lightly wooded with acacia and olive trees
Right: Zebras in the open grasslands of Meriti Valley, Mburo

Klipspringers, delicate antelopes with large ears marked inside with black and white, are found on the granite outcrops – and only there – pronking languidly in pairs across precipitous slabs of rock with the gentle touch of a pianist's hands on the keys. The churring and wing-snapping of the small agile camaroptera can be heard and their olive wings glimpsed as they manoeuvre erratically through the tangles of thorns and creepers that are their habitat. Shoebills (no longer called storks as they are considered to be more closely related to pelicans), and splay-toed sitatunga antelopes hide in the deepest retreats of swamps where little else ventures. Zebras congregate in the open valleys where their preferred short grasses are found, at least after the coarser material has been mown and trampled down by buffaloes or cattle. Leopards, of which there are many in Mburo even if they are rarely seen, hide quietly where there is cover for ambush, or rest in trees or on rock lookouts.

For every creature and plant there is a place in the numerous interconnected ecosystems of Mburo. Plants are the key component of the physical structure of these habitats, providing living spaces and microclimates for other plants and the creatures that live in, on and off them. The animals modify the vegetation through their grazing and browsing and nest-building, and change the terrain with their burrowing and wallowing and scraping, all contributing to the creation of more niches for more species.

In 1988, seven years after the aerial survey that had flown me over Mburo and stirred my attraction to the land and the tableau in which people, cattle and wildlife apparently coexisted in harmony, I returned to Uganda. I had spent the intervening years travelling and working in other parts of Africa, gathering a range of perspectives on conservation.

On this occasion I had the chance to navigate the hills and valleys to reach the lake itself. There was no disguising the parlous state of the land in its newly created status of national park, and driving through the landscape I had flown over in 1981, things not only looked different, but felt different too. The atmosphere of peace and tranquillity was gone. The wildlife I had seen from the plane had disappeared almost entirely, the extensive, open pastoral landscape was vanishing beneath banana plantations and maize and cassava fields, and there was a rancorous hostility between the few rangers that hung on in the park and the people now living in and around it.

My fond imaginings of integrating people and nature conservation, constructed on what I had seen from the air, had no doubt been naïve and uninformed, but what I saw when I returned was without doubt a step backwards from what had been in place seven years before. The decision to turn the game reserve into a national park had clearly gone badly wrong. Although some individuals had benefited by acquiring land, and others had reclaimed land that was rightfully theirs, most of the community had lost out. From a conservation perspective, wildlife had been decimated; the open valleys and wooded hills, the pastoral landscape shared by cattle and wildlife, were turning into farmland, and the local people and their leaders were dead set against the very idea of a park. Nothing useful or positive seemed to have resulted from establishing it, and a great deal seemed to have been lost.

I had been brought back to Uganda on a short contract to investigate how fishing villages within Queen Elizabeth National Park might be accommodated without compromising conservation objectives there. I thought a visit to Mburo, though something of a busman's holiday, would be a welcome break for me and my hard-working team of survey enumerators. Of course I already knew something of the short history of the park – its establishment at the point of a gun, its development, and its rapid and hard fall – and I had persuaded my team that, as budding conservation professionals, they would find a visit interesting.

We threaded the rugged hills to Mbarara, the historic capital of the Ankole Kingdom, and continued east on crumbling tarmac to Sanga Trading Centre,[31] a grand title for what was little more than a collection of mud-walled, *mabati*[32]-roofed *dukas*,[33] the ubiquitous shops of rural and small-town Uganda. All seemed to sell the same dusty bars of orange

---

31 The word '*sanga*' is used to describe cattle breeds derived by crossing two species of cattle, of which the long-horned Ankole are perhaps the best-known example. Sanga Trading Centre was not named after the long-horned sangas, as might be imagined, but after a man-eating lion with a distinctive mane that had been known to roam the area when the trading centre was becoming established – a nice example of the fact that things are not always as they seem.

32 These corrugated iron sheets are the roofing of choice in many parts of Africa when people replace traditional grass thatch. It is cheap, quick and easy to roof with *mabati* though they turn houses into ovens during the day and fridges at night, and when it rains there is such a cacophony that all conversation must stop. I have spent hours in meetings halted by the roar of rain thrashing the metal roof above as though we were part of some art-inspired installation representing universal chaos.

33 *Duka* is the KiSwahili word for the small shops that line the streets of the towns and villages of East Africa. It is a corruption of *dukkan*, the Arabic word for shop.

and blue laundry soap, plastic washing bowls, hessian sacks of broken rice and dry beans, jerrycans of paraffin and crude *waragi*[34] standing side by side on packed earth floors. Rough chopping logs covered in bone chips and gristle marked the butchers' shops, with lumps of meat and clumps of blue buzzing flies hanging from their window frames. Low dark doors showed the houses where banana beer was served in plastic mugs and tobacco was sold in papery twists of maize cob leaves.

Heading up the narrow track that led out of the village, we were soon traversing hillsides covered in eye-high grasses and scattered acacia trees. Barriers of cut acacia bush, viciously clawed capparis bushes scrambling through them, or stumpy live euphorbia hedges marked out new land claims.[35] Banana plantations, the young plants looking naked and vulnerable in the cleared fields, were just getting established. Before too long they would become a major feature of the landscape, but at the time they seemed lost in the bush.[36]

We crossed grassy valleys grazed down to a lawn-like appearance, studded with the eccentric forms of termite mounds, each bearing a shaggy crown of creepers and spiny bushes, cool green candelabra trees pushing through them. Standing water glinted in valleys where the pungent smells of crushed swamp grasses and the sticky sound of our tyres running over the wet mud warned us that we could not attempt to cross there. Dull grey ibis, jazzed with iridescent green wing bars and a crimson streak on their down-curved bills, would pace anxiously in front of us before reluctantly and laboriously taking to the air with mournful, ironic cries.

We frequently lost our way as tracks petered out, rubbed away by wide causeways trampled by cattle. A scurry of young boys would soon arrive to peer at us through the bushes, averting their gaze if we looked at them, suspicious of the large white Land Rover and its white driver. Reassured by the friendly greetings of my colleagues, they were soon

---

34 British officers gave waragi – a corruption of 'war gin' – to their African troops and, no doubt, to the warriors of Buganda who joined them in the first campaigns to suppress the Bunyoro and other neighbouring tribes. My first taste of waragi, sloshed liberally over a fruit salad of pineapple, mango, pawpaw and sweet banana, was described to me as 'African science'. It was made by distilling *tonto*, a rather sludgy beer made by trampling bananas in a wooden vat and burying the mush to ferment. Basic water-cooled stills made from oil drums are used to distil the spirit. The danger of poorly distilled bush waragi led a company to be licenses to re-distil and package safe, clean 'triple distilled' Uganda Waragi.

35 Live hedges are commonly used to demarcate land and protect fields from wild animals. In the semi-arid Ankole region, succulent *Euphorbia tirucali*, with tendril-like branches that give off a noxious white sap if broken, is the popular choice. Left to their own devices they grow into trees but planted in rows they create dense barriers.

36 The practice of creating acronyms for projects can be surprisingly effective in building the presence and identity of projects and programmes. Zimbabwe's unwieldy Communal Areas Management Programme for Indigenous Resources became CAMPFIRE, under which name it gained worldwide recognition. The similar Luangwa Integrated Resource Development Project (LIRDP), by comparison, is virtually unknown. USAID took the practice to great lengths in Uganda presenting APE (Action Program for the Environment), followed by COBRA – Conservation of Bio-diverse Resource Areas. I was pleased to propose EPIC – Evolving Partnerships for Integrated Conservation. The expansion of banana plantations into fertile valley heads and across gentle slopes removed grazing essential during extended dry seasons, reduced wildlife habitat, and changed the look and feel of the wooded savannah of the area. Arthur, Moses and I privately labelled our concern over the spread of bananas and our interest in finding ways to discourage it, BOB, our acronym for 'bugger off bananas'.

running ahead of us, long elegant legs striding, graceful arms gesturing to point the way as we swerved and swung between the trees. Where are we going, they ask. Mburo Park, we reply. Our answer is received in silence with dubious shrugs and small smiles.

The strong and abiding interest of the Banyankole people in Mburo dates back to the 15th century. Mburo was the heartland of Nkore,[37] the kingdom of the Banyankole, consolidated and extended through British influence at the end of the 19th century but abolished in 1967 when Uganda became a republic. Despite the importance of Nkore, its rulers had not always been able to retain control over Mburo. Only a few decades prior to the arrival of the Arab and European explorers and traders that would soon change everything, King Ntale took Mburo back from the weakening grasp of its northern neighbour, the Bunyoro kingdom.

The establishment of Lake Mburo National Park in 1983 can be considered the last and perhaps most disruptive manifestation of foreign interest in the lands of Mburo, certainly the most damaging in terms of the relationship between the Bahima – the Banyankole pastoralists[38] – and their Beautiful Land.

Mburo lies within the interlacustrine zone – the land between the lakes. Lake Victoria lies to the east, Lake Tanganyika to the south, and Lakes Edward, George and Albert to the west; to the north lies Kyoga, through which the Nile flows at the start of its slow meander to the Mediterranean Sea. The lands between the lakes are dramatic. Volcanoes smoke and spew lava; escarpments fall away to rivers and plains; mountains bearing permanent glaciers rise to over 5,000 metres, pushed up by the shifting tectonic plates of the Rift Valley.

Forests harbour gorillas and chimpanzees, grasslands support herds of elephants and buffaloes, and crocodiles and hippopotamus crowd the rivers and lakes. The land between the lakes is also famous for its cattle, the long-horned Ankole cattle, named, perhaps, after the place where the most beautiful are found. Ankole cattle are the especial inheritance, possession and passion of several of the tribes who live around and between the lakes. Of these, though, it is the passion of the Bahima that is so strong and so exacting that it

---

37 Nkore, previously Kaaro Karungi, was the name of the kingdom of the Banyankore people, speaking the language of Runyankore. Nkore became anglicized to Ankole, and with British support became the expanded Ankole Kingdom. I will use 'Nkore' when writing about the pre-colonial period, and 'Ankole' for the colonial and post-colonial period. The people of Nkore/Ankole will be called the Banyankole throughout.

38 The Banyankole people were divided into two distinct groups or classes, the Bahima pastoralists and the Bairu farmers, united by a common language, a common ruling clan or family, the Bahinda, and a king, the Omugabe. Their separation was geographical, social and economic. Enforced by powerful social norms, including food taboos and restrictions on cattle ownership, the separation was so effective that there was little interaction between the two except at the court of the king. Some elements of this structure remain today, but the lines dividing Bahima and Bairu have to a great extent eroded.

actually defines them as a people. It cements their bond to their homeland and has led them to create, in the Ankole breed, creatures of rare beauty.

The fall of the park

After some hours of difficult driving, and only with the help of our young guides, we arrived at the park headquarters, by which time we were smeared and speckled with the black mud of the treacherous valleys. The buildings had been constructed in 1987 on a gentle slope running down to the northern tip of the lake the year before our visit, and was indeed still under construction. On a ridge above I could make out the lines of a ruin, the roofless remains of the original headquarters buildings. The wardens and rangers that had proudly established that solid base for the new park just three years earlier had fled the angry horde that accompanied the rebel army which would go on to establish the new government in 1986, displacing Milton Obote and installing President Museveni, who at the time of writing is still president. Determined to destroy the park, this crowd looted the buildings, a practical as well as symbolic act and their first step towards reclaiming the stolen lands.

It was something of a miracle that the entire park was not irremediably lost right then, and that a small presence of rangers had returned. The distinctly more modest headquarters was constructed in a clearing halfway down Rwonyo Hill, well below the ruins on the hilltop. Perhaps the warden assigned to fly the tattered, disfavoured and

The original park headquarters on Rwonyo Hill, ransacked in 1985, renovated as an Interpretation Centre in 1993

discredited flag of the Uganda National Parks had felt it wise to keep a lower profile. Or perhaps he simply found it more convenient to be closer to the lake where he and his men collected their water and got their daily allowance of fish on which they depended.

Ducking out from under the low thatch of a row of ramshackle huts, a young man in khaki rags appeared. He greeted us warmly, excited that visitors had reached this embattled bastion of Ugandan conservation. 'You are almost the first visitors to arrive since the park was lost,' he told us.

The reconstructed headquarters was little more than a camp, guarded by this man alone. It was all very different from Queen Elizabeth National Park, where long lines of brick houses bustled with rangers, porters, workers and their families, and where there was electricity and pumped water, workshops, stores, canteens, shops and even a hostel and a hotel. Here there was almost nothing and no one. The ranger, it turned out, was not actually a ranger at all, but the gate clerk.

'I had to remain here, alone,' he explained. 'There is only one vehicle, and we have little fuel, so we drive it less often. I guard the camp while the warden and rangers go to Lyantonde for the market and maybe even to find some beer.'

Despite Mburo's obvious shortcomings, our new friend was determined that our visit should approximate what was expected from a visit to a park. He showed us to a rough area close to the lake, grandly called Campsite 2. Here we could pitch our tents. We could see where the rangers slithered down the bank to the lake to dip their jerry cans into the thick green water. If the warden had fuel, he would reverse his battered Landcruiser to the edge of the lake to fill an oil drum and drive it to the camp, water slopping and draining in torrents from the back.

We camped that night at the lake's edge near a path in the mud worn a foot deep where hippos climbed out to spend the night grazing. As I lay down to sleep I could feel their deep, yawning roars vibrating through my body. Made hesitant by our presence they grumbled amongst themselves, whooshing and pushing in the shallows before finding their way to land further down the shore. In the morning we found the splatter of their liquid dung, spread by rotating their short tails like propellers, all around our tents. We bathed, watchfully, in the small bay, the cool morning air raising a thin mist through which the sun streamed as it rose above the ridge behind us.

After breakfast – dry tea[39] and bread – I sat at the edge of the lake, looking across the still water to the hills. Pairs of fish eagles, their white and chestnut breasts glowing, sat like married couples looking out from their perches in the ramparts of ancient fig trees, peering fixedly across the margin of green and yellow papyrus for signs of their breakfast.[40] Their raucous cries rang up and down the lake. The wheezing, horsey laughter of the hippos, back

---

39 Dry tea is a wonderful bit of Ugandan English meaning tea made without milk. Fortunately it does allow the use of water, and of course lots of sugar.

40 Like some other eagles, including the American bald eagle, African fish eagles mate for life, so they really can be thought of as married couples.

in the lake again after their night on land, rolled across the water. Pied kingfishers hovered and dropped into the matt mirror of the lake. A bushbuck, delicately pointing her hooves, stepped through the herbs and forbs on her way to drink. It was magical and perfect and tranquil. All seemed safe, secure and timeless in this corner of the park.

But what we had seen on our drive the day before told a different story.

Later that morning, as we navigated unmarked tracks guided by our new friend, we heard the recent story of the park, at least the story as he knew it. He bemoaned the lack of wildlife and the countless cattle that trampled the wide, irregular tracks we followed. As we drove we came across herds, sometimes grazing alone, sometimes accompanied by thin, straight men wearing striped cloths wrapped around their waists. Bare-chested, they propped themselves against their peeled white sticks, the ubiquitous herders' tool, to regard us warily as we bumped past.

Some carried automatic rifles, and our guide, shocked still at the wrongness of it, told us, 'They cannot give them up, they refuse. Even in a national park they refuse!'

Our greetings and friendly waves were not returned. We were in Mburo now, where an uneasy truce held between pastoralists and rangers. Our Land Rover associated us with the park and perhaps with other more distant authorities; perhaps we held some influence and would intervene to their detriment. They knew about the arbitrary and opaque workings of government, and watched us with worry and suspicion.

We were led by way of valleys, skirting lakes and swamps and thickets, to the top of Kazuma Hill from where we could look out across a wide valley. The rains had washed the air clear. To the south, low hills lay on the horizon, while immediately in front and below stretched an expanse of floating reed and papyrus swamp. Three small lakes reflecting the blue sky nestled in the wetland. The one closest to us was being used to water cattle, and as we watched, herd followed herd through the bush to drink at doughnut-shaped mud troughs. Men with buckets replenished them from wells dug into the swampy lake edge. Several animals would crowd in to drink together, jostling for a place, manoeuvring their heavy horned heads to bring their mouths to the water as the men scooped more up and poured it gently into the troughs.

We could hear the 'thunk thunk 'of their horns knocking together and cries of encouragement from the men. We listened to the thudding of hard hooves on soft earth as the animals broke into a plunging gallop on smelling the water, and the whoops of the herders chasing them down to the lake. We could see over a great swathe of the park from our vantage point – but on scanning the land with our binoculars not a single wild animal did we see that morning.

To the east we could make out the southern tip of Lake Katchera. A major body of water over 20 kilometres long, it had once formed the eastern boundary of the park, but no longer. When established, Mburo, with the support of the local government, the commitment of national government, and the backing of the international conservation community, had covered 780 square kilometres. Three years later it had been reduced

Lake Kazuma viewed from the hills above

by almost two thirds, the land sliced off to resettle people displaced by the six-year civil war, to be handed out to individuals, and to revert to extensive private and government ranches. Without any interest or backing, the area that remained park languished. Even Uganda National Parks, responsible for its creation and for its continued protection, seemed embarrassed by Mburo and gave it as wide a berth as they could.

The creation of a national park cannot be taken lightly; it commits the nation to protect it – in theory at least – in perpetuity, so it is a decision taken at the highest levels of government. The creation of a national park attracts the attention, approval, and sometimes even the assistance of other governments and development agencies. International conventions encourage and even commit governments to protect biodiversity,[41] most commonly by establishing protected areas, and a plethora of charities exist to help. Though a range

---

41 A number of international movements to achieve the conservation of wildlife have been pursued since the Convention Relative to the Preservation of Fauna and Flora in the Natural State, otherwise known as the London Convention, was agreed in 1933 by several colonial administrations. The most important today, the Convention on Biodiversity, established in 1992 and signed by 150 nations, requires biodiversity to be conserved and sustainably used, and the benefits of its use shared equitably. The United Nations Sustainable Development Goals require governments to undertake activities that will deliver 17 goals to achieve sustainable development by 2030. Goal 15 – Life on Land – requires governments to protect, restore and promote sustainable use of terrestrial ecosystems, sustainably manage forests, combat desertification, and halt and reverse land degradation and halt biodiversity loss.

of types of protected area exists, designed for different objectives and with different requirements, rules and regulations, it is the national parks that are considered the most prestigious.

National parks are understood as an enduring bequest to the future, and closing one down, reducing its size, or allowing it to decline does not go unnoticed in the corridors of international power and influence. President Museveni, the newly installed head of Uganda's government, having come to power through force of arms, was anxious to be seen as sound and responsible by the community of nations. For him to reduce, as he did, the size of Lake Mburo National Park as one of his first public acts there were compelling reasons. Pushed by demands to redress the wrongs done when the park was created, this is what he did, in 1986. He had committed himself to this whilst still fighting in the bush, and he needed to deliver the land promised to his warriors and supporters.

Even after its reduction in size, the intensive cattle grazing and unchecked hunting reduced the wildlife numbers, the expansion of farming cleared the woodlands and thickets, and the unregulated fishing competed with the fish eagles and pelicans and killed the crocodiles. The park's rangers, without political backing, were powerless to intervene in the face of local resistance to the very idea of the park. Settlements were growing fast, and banana plantations were extending down the valleys and across the hills. Schools and churches were constructed as communities bedded in, demonstrating their permanence and strengthening resistance to any who might attempt to remove them. The development agencies that arrived in Uganda after the civil war were sinking boreholes and installing water pumps across the area cut from the park, but also within the area that remained park.

A weekly market located close to the lake was granted a licence by the local Resistance Council,[42] attracting fishermen to trade salted and sun-dried tilapia, smoky black coils of catfish skewered on sticks, and the prehistoric lungfish, still alive, secreting slime, heaving and churning in papyrus baskets. Then farmers came with vegetables and chickens to sell or barter. Hawkers arrived on the backs of rusty pick-ups or by bicycle or on foot carrying bolts of printed cloth, bundles of multicoloured sandals and bales of second-hand clothes. Perhaps it was reasonable to facilitate the everyday needs of the people who found themselves living in the park – the quotidian life of rural communities has to go on – but the level and nature of the activities of business as usual, and their steadily increasing intensity and range, was now undermining the future of the park and its wildlife. The number of cattle grazing in the park alone went beyond what the area could sustain, even had it been a ranch and the only interest the production of meat and milk.

But through the 1980s Mburo, the national park, retained the purpose, on paper at least, of protecting nature.

---

42 As Museveni's rebel National Resistance Army acquired territory, a system of administration and political representation was put in place. Resistance Committees (RCs) were established from District (RC5) to village (RC1) levels. Committee members were elected and initially owed allegiance to the new government of the National Resistance Movement. In time these committees were renamed Local Councils.

Left: Great white pelicans Pelecanus onocrotalus *roosting at Lake Mburo in 1988. Competition from intensive fishing led to their decline and disappearance from Mburo*

Upper right: Tilapia fish from Lake Mburo, salted and sun dried for market

Right: Gill nets are supported with floats made from Ambatch, the balsa wood tree.

It was clear to me that the park was in difficulty and that life for the rangers, reviled by their neighbours, was almost impossible. The Bahima pastoralists who in 1985 had occupied Mburo under the wings of Museveni's advancing army were determined that it should never again be given over to wild animals. They identified Mburo's open, undeveloped rangelands and its herds of wild animals as the inspiration of this incomprehensible madness; if those were the values that had attracted the interest of government, them they should make the land as unattractive for conservation as quickly and completely as possible, even if this meant compromising some of their own interests.

Individual Bahima herders had led the charge, bringing in their cattle to occupy the land and swelling their numbers with others' cattle, especially those belonging to senior army officers. Though hunters need little encouragement, an invitation was extended throughout the region, and bands of men soon arrived with spears, nets and dogs. Though the Bahima do not eat game meat, most Ugandans do, so the strong demand and the inability of the rangers to mount any level of protection meant that the herds of impala, topi and buffalo, the scurrying families of warthogs and the pods of hippo were quickly decimated. Only zebra and waterbuck remained to get fat[43]

---

43 For reasons that are not clear, none of the local tribes are keen to hunt or eat zebra though, unlike the waterbuck, which is also avoided, zebra meat is good to eat.

Many farmers had been evicted when the park had been created, and they had returned, to be joined by others who saw a chance to acquire land. Family, friends, even enemies were called to join the scramble, and within a few short months much of the land had been parcelled out and settled. The area retained as park lacked an agreed boundary or any active commitment to its protection. It too was settled, and it was hard to tell that Mburo remained a protected area. The valleys and hillsides were full of cattle; trees, great and small, were burned for charcoal; and land was steadily cleared for farming. The tracks that connected the communities bustled with activity while the wild animals grew fewer and hid themselves in distant corners.

Having observed these changes, we made our way back to the Land Rover we had left at the bottom of the hill. But then our spirits, rather depressed by what we had seen from the look-out, were raised, along with our adrenaline levels, by a buffalo, a lone male, grumpy and un-predictable. He appeared suddenly before us as we made our way down the hill. Alarmed by our voices he snorted, stamped and stared at us before crashing off through the thick wood-land. Our spirits were further raised, and our nerves calmed, by a flock of red-faced love-birds, flashing like tropical fruits as they bombed through the trees, screeching and whistling.

Apparently, despite the efforts to destroy it and notwithstanding the overwhelming presence of cattle and the harrying of the wildlife, the park retained at least some of the values for which it had been established. Nature is resilient, and in the face of a history of destructive policies and practices and the most recent efforts to eradicate the wildlife, it was clearly hanging on.

As we drove back to our campsite we passed a homestead, the family coming out to watch us go past. Bahima households were scattered everywhere across the park but, as mentioned earlier, they were so natural in their construction that they were hard to spot, seeming to be nothing more than stacks of hay or the nest of some giant bird.

Our guide offered to introduce us to the herders. 'You can talk to them,' he suggested hopefully, 'about why they should not be here, and tell them to stop carrying their guns. Tell them the park is for the future children.'

I declined, doubting that any discussion would be constructive. The differences between the park and the pastoralists were not going to be resolved by strangers in a Land Rover dropping by for a cup of tea.[44] The differences were too complex and too charged, and the conflict still far too raw.

---

44 A visit to a Bahima homestead is almost certain to entail the offer of a cup of tea, and not just at teatime. The tea is very different to that which might be offered in an English drawing room in fine china, or an English kitchen in mugs. Referred to as 'African tea', to distinguish it from either 'English tea' or 'dry tea', it is thick, strong and sweet, made by boiling tea leaves in milk and water, sugar and spices, and is exceedingly delicious when made from the rich milk of Ankole cows.

The causes of conflict

When things go wrong it is always comforting to point fingers and identify culprits, and some have been identified as responsible for what occurred at Mburo. President Milton Obote stands at the head of the queue. He had two cracks at ruling Uganda, but his unswerving adherence to his own brand of African socialism won him few friends and many enemies. Both of his tenures as president ended in his ejection, first by Amin, ushering in a decade of horror and decay, and then by Yoweri Kaguta Museveni, who has ruled ever since.

Many argue that Obote's determination to establish Mburo as a national park was politically motivated. Few Bahima voted for him during the obviously rigged election that brought him to power in 1980. He is reported to have said that as so few from the area had voted for him it must be that no people lived there, and if there were no people in the area, then there could be no objection to establishing a national park in it.

Driving the Bahima out of Mburo, a land of cultural as well as economic importance, would be their punishment for voting for his opponent, Museveni. Obote's Minister for Security, Rwakasisi, purposefully blurred the distinction between the Ugandan Bahima and the Tutsi refugees from Rwanda living in the same area, his harassment of both approaching ethnic cleansing.[45] During this time, many Bahima lost their lands and their cattle. Some lost their lives. The creation of the park was seen to be part of this persecution, and the Bahima still consider the park as a stick conceived to beat them with.

Obote was not the only player behind the creation of Lake Mburo National Park, nor the only influence on how it was to be achieved. Sitting over the campfire that evening by the side of Lake Mburo, listening to the descending three-note call of fiery-necked nightjars ringing around us and watching the moonlight's glimmer on the moving surface of the lake, I had to accept that I too had been a player in the story. Though a minor actor, I would, in retrospect, have rather had no part in it. The counts of cattle made during the aerial survey I had been part of had revealed a fourfold increase in their number. We had also counted a not inconsiderable number of wild animals, estimating over 6,000 'plains game' in the reserve.[46] These figures gave to those wanting to strengthen the protection of Mburo the data to argue that the area had important conservation values.

---

45 Many Rwandan refugees settled in Uganda following the conflict between Hutu and Tutsi, the two main ethnic groups of Rwanda, in 1959. Obote attempted to undermine support for Museveni and his insurgency by branding him a Rwandan. The Ankole and Rwandan people have many similarities, and Bahima and Tutsi both keep long-horned cattle. Museveni, from a Bahima family, was vulnerable to this attack. The Bahima suffered badly under Obote's rule and there is evidence of pogroms against them and the largely Tutsi Rwandan refugees.

46 We carried out our count during the rainy season during which period much of the wildlife disperses out of the area around Lake Mburo and therefore leaves the nature reserve and would not have been counted. It is likely, therefore, that the total wildlife population for the area was considerably higher than this.

That Uganda's last remaining impalas lived in the area was a further, and emotive, argument for creating a national park.[47] Though impala is one of Africa's most common antelopes, Lake Mburo represents the most north-westerly extent of their range. Whether this can be considered a sufficiently strong argument for making Mburo a conservation priority can be questioned, but the Obote government clearly gained international approval for protecting the wildlife-rich area. Uganda was commended for creating this new national park.[48]

Severe conflict between the conservation authorities and the people who lived in and grazed Mburo was inevitable. It arose almost immediately, and though the steps that were taken can be seen as the result of a commendable wish to protect Uganda's natural heritage, this cannot excuse what followed. The story I tell here is primarily that of the Bahima, but theirs was not the only story of loss. In short order, following the official declaration of the park, measures were taken to evict everyone living there. Cattle-keepers, subsistence farmers, fisherfolk, traders, shopkeepers, bicycle repair men and all were thrown from their homes and erased from the landscape. Even the wealthy individuals and institutions that had acquired ranches in the area could not resist, and their lands were confiscated.

I make no apology for focusing on the losses experienced by the Bahima in particular. Their story provides a powerful example of the negative impacts of conservation as it was – and too often still is – practised, combined with our failure to engage with the interests and values of local people, and our consequent failure to win real support for protecting nature. The story of Mburo is not unique; many parks and reserves have similar histories of increasing control by government and steadily ever more imposed separation of the people from their lands. The circumstances of Mburo's rise and fall and its impacts on the Bahima are relevant not only to Uganda but to many other countries. And my own journey, for good or bad, has been bound up with the story of the Bahima and Mburo.

A history of connection, conflict and loss

Conflict over Mburo and competition for its lands, waters and resources did not begin with the creation of the national park in 1983, or even with the creation of the game reserve 20 years before that. There was ancient conflict with neighbouring tribes for control of the

---

47 Uganda's other populations of impala were in Katonga Wildlife Reserve to the north of Mburo and in a corner of Karimoja in the east of the country. Both populations were believed to be extinct, though impala are found today in Katonga, which is to be made a national park.

48 Jonathon Kingdon, writing in the *Oryx* journal in 1985, praised the creation of the park, describing it as significant and welcome evidence of the resuscitation of the country's commitment to conservation and makes a plea for international support for the park.
'Lake Mburo – a new national park in Africa', Kingdon, J. (1985) *Oryx*, Vol 19, Issue 1, pp. 7–10.

land, but within the Nkore kingdom itself there was also competition between pastoralists and farmers. Fig and mango trees mark the sites of settlements throughout the area, especially along the shores of Mburo's several lakes, demonstrating that the lakes and wetlands had long supported fishing communities and that farming had also been carried out. Settled communities existed across Nkore, some short-lived, others more permanent, providing food and shelter to travellers moving between the ancient kingdoms of the region through the expansive and inhospitable savannah woodlands and grasslands. In the second half of the 19th century Arab and European explorers and traders began to arrive, and they were able to journey along well-travelled routes between villages large enough to supply their caravans comprising hundreds of porters, soldiers and hangers-on.

The potential for competition for land between the pastoralists and farmers of Nkore was constrained by the social, economic and political organization of the kingdom. The particular conflict that played out during the rise and fall of Lake Mburo National Park and which continues to taint the relationship between park managers and their neighbours today, began with the arrival of British colonial officers at the end of the 19th century. Pursuing the interests of Empire, they demanded changes that disturbed the delicate balance between pastoral and farmer groups and their rulers. Events were set in motion that almost a century later permitted the creation of the park and the punitive way in which it was created.

The mixing of oral traditions and history make for a complicated and contested story, but it is generally agreed that the peoples of the Great Lakes region were part of the great Empire of Kitara, also known as the Empire of Light. It covered most of modern-day Uganda, large parts of northern Tanzania, western Kenya and eastern Congo until it began to fragment in the 14th century. Most of the peoples who occupy this region today claim a connection to the Bachwezi, remembered as the mythical demi-god rulers of Kitara.

As a product of this common history, the current peoples of the region share a number of social, political and cultural features.[49] When Kitara finally fell, smaller entities formed. Within these kingdoms or principalities, the people were divided into classes, generally a ruling clan, a pastoralist group and a farming group. Though united by a king and a common language, the classes would live different and sometimes quite separate existences. The people of these kingdoms also shared something else: the ownership of the long-horned cattle. How the Bahima of Nkore came to own these beautiful creatures lies at the centre of their story.

Wamala, the King of the Bachwezi, determined to kill himself. His most beloved cow had died suddenly, and he could not live without her. Scoffed at for such an extreme response, even for a Bachwezi – to give up the life of a god for the life of a cow – Wamala was persuaded to give up the idea. But when his aunt then mocked him for failing to keep

---

49 Most notable of these are the existence of centralized power structures under a hereditary ruler, drums acting as the symbols of power and unity within the state, and separate groups or classes within a single people. Not all these characteristics are held by all the peoples of the region.

his word, he cried out that he could no longer live in a world where he was so humiliated. His life dishonoured, the world now forever tarnished was no longer fit for a god. His fellow Bachwezi agreed with him and agreed to follow him. So he led them below the waters of a great lake, never to be seen again.[50]

Ruhinda, one of Wamala's sons, had however taken an interest in the affairs of humans. Rather than follow his father he determined to remain behind, becoming the first of the kings of Nkore. He kept his father's herds and, gifting them to the Bahima, charged them to look after and love the cattle as the Bachwezi had.

Though the mix of myth and history is confusing, this is all believed to have happened towards the end of the 15th century. That was when the dynasty of the kings of Nkore started, and it ended with the abdication of its last king, Gasyonga II, in 1967, when Uganda was declared a republic.

The kingdoms that had arisen following the vanishing of the Bachwezi were smaller and weaker than the Empire of Kitara, but retained some of its characteristics. For 400 years the people of these kingdoms competed, raiding each other's cattle, seizing each other's territory, quarrelling and trading. For centuries the Kingdom of Nkore was sandwiched between more powerful kingdoms and was hard pushed to retain control of its richest grazing lands. For decades Bunyoro, the most powerful of the kingdoms, held the heartland of Nkore, the famed hills and valleys of Nshara where the royal herds should graze.[51] But as the fortunes of raiding, occupation and alliance swung back and forth, and as kings and princes rose and fell, ruled and died, the grazing lands, the cattle and the power changed hands.

These local power plays could be fierce, but their impacts on the land and its peoples was about to be eclipsed. By the middle of the 19th century the interests of people we can truly call outsiders were about to make their presence felt. The time of British imperialism in Africa was beginning, and a different world with different values would soon hold sway. Foreign interests and a distant government would determine how the valleys and hills of Mburo would be used, who would control them, and even what they would mean. But as foreign interests and values began to change the meaning of the land, the Bahima, to whom Mburo meant the most, began to resist.

---

50 Lake Wamala lies to the east of Ankole and, as its name suggests, is closely associated with the disappearance of King Wamala and the Bachwezi. The lake is believed to be the one that they descended into and is an important site for prayers and rituals for many peoples in the region. The idea of a whole people descending into a lake is an intriguing one. A tale is also told of mythical warriors being led into a lake in Ireland.

51 The Bahima name their animals depending on aspects of their appearance. *Nshara* is the name for one with small dangling horns. It is surprising that this animal, not generally considered by Bahima to be beautiful, should have been so favoured by the king, who kept a herd of Nshara cattle. It is not clear whether this is the origin of the name given to the Nshara area.

In the 1860s, the country we know as Uganda today was known to only a small number of outsiders. By the mid-1840s Arab traders and slavers had reached the Kingdom of Buganda, established relations with its king, and opened trade routes. For Europeans, however, these lands remained hidden in the remote interior of the Dark Continent. If considered at all, it was as a possible location of the mysterious Mountains of the Moon, mythical source of the Nile, or perhaps as the long-sought final resting place of Prester John.[52] But as Arab traders pushed further into the interior from their coastal forts, and slaves and ivory flowed out in ever more horrible quantities, Europe began to pay attention to eastern Africa.

Indeed, Europe was beginning to cast a hungry eye on Africa in general. Despite its contact with coastal communities since the 16th century, the establishment of a few trading towns and ports, and the Dutch settlement on the southern tip of Africa, knowledge of the interior was limited. So the African empires, nations and tribal groupings quietly continued their use of the land and its resources, and throughout the continent their leaders ruled their peoples almost unchallenged, safe from the attentions of the west.

Yet within a few short decades of the explorations of Livingstone, Burton, Speke and Stanley in the mid-19th century, the entire continent with its peoples, excepting Abyssinia, was subjugated by European powers. The Dark Continent, so named because maps of Africa were almost entirely blank, was suddenly and traumatically transfigured into 'darkest Africa' – a graphic depiction of the horrors visited on this vast continent. Though the ravages of the slave trade perpetrated by Europeans in west Africa over three centuries had wrought havoc in the interior, depopulating and destabilizing whole regions, the impact of the European colonization – in East Africa, at least, ironically stimulated by the purported desire to halt slaving by Arabs – would be as great.[53]

It would be fair to say that British interests were not driven entirely, at least initially, by the desire to acquire land, resources and markets for their products while depriving rival nations, especially the French, of the chance to acquire the same. Instead, having been an enthusiastic participant in the slave trade for centuries, the British government bowed to pressure and abolished it in 1833, and then used its navy to suppress the trade by others. In East Africa, though, the taking and trading of slaves continued under the Sultanate of Oman and Zanzibar.

The attention of missionary groups was drawn to this, especially after David Livingstone's journey into the interior in the 1860s and his reports of the horrors he saw

---

52 An ancient map drawn by the Greek scholar Ptolemy showed the Nile River flowing north from lakes lying at the base of a range of mountains labelled The Mountains of the Moon. Though no reliable sources existed to verify this information, reports from Arab traders suggested there were great lakes lying somewhere in the interior. Prester John is a legendary king of a Christian nation lost in pagan lands. Though earlier legends placed him in Asia, later stories linked his kingdom to Abyssinia, today's Ethiopia, a Christian nation since the 4th century.

53 The nightmares created by Europe's insatiable demand for resources were captured by Joseph Conrad's disturbing masterpiece, *The Heart of Darkness*. Published in 1899, not much more than a decade after the annexation of the Congo, it tells of a journey up the Congo River in search of Mr Kurtz, an ivory trader lost in the interior. The story was heavily based on reports of atrocities and the barbarous treatment of the native people by traders and bureaucrats that were seeping out of the Congo region.

there. While religious groups were keen to pursue humanitarian concerns and began organizing themselves to enter the region, the British government began to worry that European rivals might steal a march on them, and begin annexing territories. In the 1870s and 80s the unseemly scramble for Africa began, and it was two clearly laudable interests – suppressing the slave trade and solving the mystery of the source of the Nile – that provided convenient points of departure for the British to extend their influence.[54]

The source of the Nile remained one of the outstanding trophies of exploration in the mid-19th century. Journeys of discovery attracted enormous public interest as well as the attention and support of the scientific community, and the illustrious members of the Royal Geographical Society were keen to win this trophy for Britain. Though the first expedition, undertaken by Richard Burton and John Speke in 1857 and 1858, ended in failure, they were at least able to agree on the existence of two great lakes deep in the interior. Unfortunately, they could not agree which, if either, was the source of the Nile, and fell out badly over the question.

Speke set out again two years later to resolve the dispute. Travelling round the western edge of the lake he had named after Queen Victoria two years earlier, he passed through the lands of Karagwe, in so doing becoming the first European to describe long-horned cattle. Continuing to the lands of the Baganda on the northern shores of the lake, Speke correctly identified the great river he found flowing north out of the lake as the White Nile. He followed it all the way to Khartoum in Sudan, where it joined the Blue Nile flowing from the mountains of Ethiopia.

The report he submitted to the Royal Geographical Society detailing his discoveries was lamentably muddled, and doubting voices were raised, including that of his old adversary, Burton. In an accident that would be considered outlandish in a novel, on the day before the meeting at which he was to clarify his findings and claim the great prize he accidentally shot himself dead. It was ten years before Henry Stanley finally proved that Lake Victoria was the source of the Nile by circumnavigating it and proving that Lake Tanganyika, the second contender, could not be the source.

In 1877 Scottish Protestant missionaries arrived in Buganda. Catholic White Fathers arrived two years later, and both began converting the Baganda people with enthusiasm. Relations between the missionaries and their converts were soon strained by the inevitable conflicts between their religions and the respective interests of Britain and France. The Kabaka of Buganda, the supreme ruler, uncomfortable after five years of tolerating the disruptive influences of the missionaries that were eroding his authority, pushed back ferociously. First, he arranged for James Hannington, the first Bishop of East Africa, and

---

54 In 1884, the European nations met in Berlin to divide the African continent between them. Britain, France and Germany had already begun to compete for control over strategic locations and to avoid having to fight each other to grab African lands, it was a lot more pleasant and efficient to simply agree who would take which parts. The Berlin Conference was very effective and Africa was divided up without spilling a drop of European blood. Africans were not invited or considered. Between 1880 and 1910, 80 per cent of the entire African continent was taken into European control.

his party to be murdered before they could enter the country in 1885. Two years later he executed 45 Christian converts, an action modelled, it seems, on his father's earlier massacre of Muslim converts. This failed to dislodge the missionaries, though, and Christian converts deposed him a year later, causing a civil war that lasted four years and cemented Christian influence in Buganda and neighbouring kingdoms.

Contrary to government policy at the time, Britain fell almost accidentally into the position of imperial conqueror in Uganda; though the British government was not keen to annex Buganda, circumstances forced its hand. The fighting between Protestant and Catholic converts was decisively settled in favour of the Protestants when Frederick Lugard arrived, employing his soldiers and his famous Maxim machine gun to end the fighting. Buganda, now in the hands of Protestant converts, leaned toward Britain, but two years later Lugard's employer, the British Imperial East Africa Company, decided to withdraw. Lugard, appalled, travelled to London to demand Buganda be annexed. The British Protectorate of Uganda was declared in 1894, three decades after Speke's first arrival in the country.

It was all well and good for Lugard and others to demand British protection for lands and peoples from Arab slavers – or worse, the French – but how was it to be paid for? New territories, the Foreign Office insisted, should be taken into the fold only if there were profits to be made by doing so. Unfortunately, however, there appeared to be little immediate prospect of Uganda producing anything useful. Worse, the distance and the ruinous terrain between the coast and the interior were a serious impediment. Early explorations had used porters, taking many lives. So reliance on porters – hardly surprisingly, uncertain and morally suspect – stimulated the construction of a railway from the coast to Uganda.

The railway earned its name, the Lunatic Express, early. In time it would allow goods to flow from the interior to the coast, but the cost of construction was astronomic, and at the time it was built there was nothing to trade.[55] The kings and princes of Uganda might be interested in the riches of Europe, but had nothing to exchange for them. There was ivory, but not enough to fund the administration of a small if charming territory in the interior, let alone the cost of building the railway. Something needed to be done, and on the basis of being in for a penny in for a pound, expansion was the given answer. The Kingdom of Buganda had been the starting point, but it was not long before neighbouring kingdoms, including Nkore, came in for attention. Its king tried to keep the British at arms' length, but more determined resistance was not possible.

---

55 *The Lunatic Express* gives a compelling account of the building of the railway. One of the most frightening times for the labourers was during the transit of Tsavo. The horrors of attacks by man-eating lions and the campaign to destroy them are the subject of the aptly titled *The Man-eaters of Tsavo* by John Henry Patterson, an engineer on the railway. Patterson finally killed the two man-eating lions, but not before they had killed as many as 135 Indian and African workers.
The Lunatic Express: An Entertainment in Imperialism, Miller, C. (1971) London: Macmillan, p. 559.
The Man-eaters of Tsavo. Patterson, J.H. (1907) London: Macmillan and Co., Limited.

Nkore was being pounded by disasters of biblical proportions. Epidemics of smallpox, rinderpest and jiggers dealt out death and destruction, triggering famine. Smallpox had probably reached Nkore with the caravans of Arab and European explorers, and by the 1890s the area was in the throes of a serious epidemic. Rinderpest, a virulent virus that swept Africa from Eritrea to the Cape, reached Nkore in 1890, decimating livestock and wildlife. Up to 90 per cent of Nkore's cattle were lost in the first wave, causing hunger and starvation, and disrupting the social fabric with lasting consequences[56] Finally, the jiggers, also called sand fleas, reached Uganda from the Congo coast, probably carried by Stanley's expedition[57] Jiggers infected feet, and though they were easy to remove there had until then been no knowledge of them – so, left to fester, they caused horrible and often fatal infections.

The responsibility for all these disasters can be laid at the feet of Europeans, but there was apparently no accompanying intention; it would be wrong to equate this with the spreading of smallpox by settlers in America to destroy its native peoples. Nonetheless these epidemics weakened Nkore, and no real resistance was offered to British demands. King Ntale V reluctantly allowed Lugard passage through his kingdom in 1891, but refused to meet him. Ten years later, Nkore was a British protectorate.

It is a remarkable fact that just four decades after the arrival of Speke, the first European to reach the region, Nkore, a proud and independent nation, was a vassal state. In sum, in 1861 Speke had skirted Nkore, providing the first accounts of the kingdom. Christian converts began arriving 20 years later, fleeing the civil war in Buganda. Henry Stanley famously swore blood brotherhood with a representative of the king in 1889 as he led the rescued Emin Pasha back to Dar es Salaam on the coast. Lugard arrived in Nkore two years later – and by 1901 Nkore was part of the British Empire!

The British were immediately attracted to Nkore and the proud and aristocratic-seeming Bahima[58] and their fabulous cattle. Their infatuation did not last long, however, as it soon became apparent that the Bahima were not interested in trade. Speke and Stanley had paid their way with beads, cloth and guns, but the traders that followed needed something

---

56 The Banyankole, badly affected by these epidemics, were unable to offer any resistance to encroaching British influence, but they survived. Other tribes were less lucky or less resilient. Losses of up to 90 per cent of the livestock and wild animals across Africa led to famine and widespread social and economic collapse in many regions. During this period, leadership failed and cannibalism as a survival mechanism was reported by early European explorers, and may be the origin of the widespread belief that Africa was full of cannibals.

57 Female jiggers, *Tunga penetrans*, the smallest species of flea measuring just 1 mm across, burrow into the skin of feet and toes after mating. The eggs mature under the skin, causing intense irritation and pain. The egg sacs are easily removed with a pin but if not removed can cause serious infections.

58 The main reason that the British considered the Bahima to be aristocratic seems to have been because they had fine noses compared to the noses of the rest of the people. The British, it seems, were rather impressed by noses.

in return for their goods. Wildlife was an early target, and it is no coincidence that the first regulations passed in Ankole in 1900, even before the protectorate treaty had been signed, were a transparent grab for a share of the ivory that was beginning to flow. Game regulations were part of the Ankole Agreement, which also introduced the first taxes. Ivory became the first money-spinner for British Uganda and for the new Ankole government. Fortunately for the British, the Bahima were not interested in wildlife and failed to understand that these first controls over the use of wildlife would soon extend to comprehensive control over the land.[59]

Ten years later things would get trickier with the introduction of the Cattle Reforms. This attempt to force the Bahima into economic production struck at the heart of their way of life and their values. They had absolutely no interest in the idea of production. Certainly, like other pastoralists, they depended on their cattle to meet their needs, but beyond that, the interests that defined them, drove them, informed their practices and social interactions, and determined their world view were the very opposite of production. To crystallize this into a single idea might be simplistic, but nonetheless the Bahima were driven by the idea of *beauty*. It was no coincidence that their name for their homeland was Kaaro Karungi, the Beautiful Land, or that their beloved cattle were bred for their beauty.[60]

This was all very frustrating for the British because, whatever they might think about beautiful things, places and people as individuals, it was the demand for wealth and production that drove them. The East Africa Company had spent mountains of cash at unsupportable levels on roads, ports and railways until Lugard's battle for Buganda effectively finished the company off. The Uganda Protectorate acquired its assets but also its debts. By the start of the new century, railway cars were waiting in the port of Kisumu, the railhead on Lake Victoria, waiting to be filled with goods, and the bureaucrats were getting desperate to have something to show for what many believed was an ill-considered decision to take on yet another colonial possession.

So the Cattle Reforms, intended to pressurize the Bahima into managing their herds for production, introduced taxes to be paid on sales of ghee and cattle. The pastoralists, seeing this as undermining their beauty-centred worldview, point-blank refused to pay. In response, the British lost little time in dismantling the institutions that supported Nkore's traditions. First, following the inauguration of King Kahaya II in 1895, the British moved to isolate him from advisors they found uncooperative. Then they arrested and exiled the

---

59 Though the Bahinda, the royal clan of the Banyankole, ate game meat and engaged in courtly hunting, and Bairu farmers hunted and ate game meat, the food taboos of the Bahima restricted them almost entirely to consuming the milk and meat of their cattle. An exception was made for the drinking of millet beer, which could be excused because it was pale and creamy and a little like milk if looked at through closed eyes.

60 Language is a difficult thing at the best of times, and translating from one language to another is fraught with errors. 'Beauty', a word that may sound so clear and easy, is the opposite, because it deals with perceptions that are not universal. *Kaaro Karungi* can be translated literally as 'the good place inhabited by people' or 'the good land'. Often described as 'the land of milk and honey', it is most generally translated as the Beautiful Land, not referring to its aesthetic beauty but to its qualities for rearing cattle.

chief advisor and dismantled the Royal Clan, the Bahinda, which supported the King and made up his court.[61]

The wholesale replacement of the King's advisors and chiefs that was thus engineered by the British would powerfully affect how the Beautiful Land would be perceived, owned and used in the future, and it paved the way for the control of Mburo to change hands. Even so, it would take some time for those empty railway cars to be filled with goods from Ankole.

The Bahima were dismayed by the threat to their role as keepers of the nation's pride, its long-horned cattle; and in this case Britain's well-tried strategy of co-opting or replacing ruling élites had disappointing results. Rather than doing as expected of them, the Bahima voted with their feet. As nomads, with mobile assets, they could simply walk away from problems – and to avoid the demands of the 1910 Cattle Reforms that is exactly what they did. This outcome, which no doubt surprised and aggravated the British administrators, proved to be a pivotal point for the Bahima, and for their relations with government too, and would in the future have a profound influence on their responses to the decisions made over Mburo and its wildlife.

The conflict over Lake Mburo National Park can be seen prefigured in this first battle of wills, which had established a pattern of resistance to change through avoidance amongst the Bahima. However, although it helped them protect their treasured long-horned cattle, it did not help them hold onto the land; left unprotected when they departed, the land was lost to them, and in the end they would lose their cattle too.

The Bahima's unwavering focus on their cattle excluded the possibility of other interests being expressed. The Beautiful Land had been dedicated to the cattle for centuries, and could be seen in no other way. With the departure of so many Bahima and with the British champing at the bit, change was inevitable. Though the government established to govern the new Kingdom of Ankole, under Kahaya, was basically a Bahima administration, it was made up of new Bahima, an educated élite which became increasingly divorced from the core values and concerns of the traditional pastoral Bahima. That élite, with the support of missionaries and Christian converts, would embrace British values. The Beautiful Land would from that point be dedicated not to beauty but to production.

British attention remained focused on filling those empty railway cars and the empty coffers of government. While rinderpest and smallpox had stressed the historic distribution of wealth and power between pastoralists and farmers, opening the way to change, the replacement of Ankole chiefs and the Bahima exodus opened the way wider. The Beautiful Land was opened to agriculture. Farmers from the crowded Kigezi highlands in the west,[62]

---

61 Amongst the kingdoms of the region, the key regalia of the kings was a drum, not a crown. Capture the drum and you capture the kingdom.

62 The Kigezi highlands are amongst the most densely inhabited farming landscapes in the world, with over 600 people living and farming on a square kilometre. The steep hills have been intensively terraced to form a landscape of great beauty and high productivity. The highlands were already overcrowded at the start of the 20th century since when, with government support, the Bakiga people have migrated to many other parts of Uganda.

farmers from within Ankole, and people who both kept long-horned cattle and farmed in traditional ways, settled in Mburo, some by invitation, some by opportunism. Fishing was encouraged, and the trade in salted and smoked fish from the lakes began.

None of this could have happened if Mburo had continued as the special preserve of the king's cattle. But the new powers in Ankole had different ideas about what should be done with good land. The British set in motion changes that placed over a third of Ankole's chieftainships in the hands of farmers. Those that now ruled the Beautiful Land had a different outlook from that of the pastoralists, and the very appearance, sounds and smells of the land began to change.

The first step towards exclusion

As villages, farms and fences spread, the new rulers of Ankole finding their control over the land slipping away, seem to have had second thoughts. The opening of Mburo to farming and its lakes to fishing also opened it to hunting. Though the Bahima were not hunters, the king and his court were. More significantly, the decision-makers of both the British Protectorate and the Ankole Government were interested in controlling the trade in ivory, rhino horn and hides; indeed, the Protectorate's first regulations dealt with the sale of hunting licences.[63] Wildlife populations had bounced back from the rinderpest epidemic and were interfering with the plans to turn Ankole into a cornucopia. Farmers were struggling to survive as crops were destroyed in the fields by elephants, buffaloes and bushpigs, while hippos and crocodiles were slowing the development of the fishing industry. Opinion had it that regulated hunting could turn the burgeoning wildlife from a problem into a solution.

It is ironic that it was King Kahaya who took the first step along the path that would culminate in the creation of Lake Mburo National Park. Sometime in the 1930s he requested game guards be posted to the area. His reasons are not clear, but it seems he was unhappy that farmers and fishermen were hunting in the ancient Ankole heartland. Using nets, spears and dogs, hunters were after small game like impala or bushbuck, not the elephants, rhinos and buffalo that earned his government revenues from the sale of hunting licences and exports of ivory and hides. Perhaps he had realised, too late, that the land itself was running through his fingers.

The Nkore system of patronage revolved around the giving and receiving of cattle, and the obsessional focus on beauty achieved a kind of passive protection of the land. People without this single-minded pursuit of beauty, people who were not prepared to sacrifice

---

63 Controls over hunting had multiple objectives, discussed in Chapter 3. As well as providing revenues through the sale of licences, they controlled how hunting was to be carried out. Trapping was considered unacceptable, and only shooting with rifles was licensed, which effectively excluded most Africans.

wealth to achieve it, could find no place in the community, and so were effectively excluded from settling on the land. They could not sustain themselves, socially or economically. For centuries this system controlled and limited the use of the land to pastoralism, maintaining the Beautiful Land as open rangeland. But now that system was well and truly broken, and the land was changing fast, and it seems that the king was irked by the open hunting by farmers and fishermen in what had once been a royal preserve.

An effect of the rinderpest epidemic that had emptied Mburo of its people and cattle as well as its wildlife was that the grasslands, no longer managed by the Bahima through fire and grazing, became bushy. This paved the way for tsetse fly numbers to explode. These biting flies are not only unpleasant for man and beast, but they also carry parasites, nagana, that infect cattle. Nagana was not always fatal, but it was a big enough threat to drive away the remaining Bahima. This second exodus left Mburo even more exposed.

Asking for game guards to restore control over the land had no doubt seemed a good idea at the time. Promoting hunting and sending hides and ivory down to Mombasa might keep the farmers and fishermen at bay for a while, but in the long run it would sharpen the conflict between the Bahima and the government. The king might have been trying to protect Mburo to retain a residue at least of a way of life, but the chapter he opened would end with the final removal of the cattle that gave meaning to the Beautiful Land.

Looking back it is easy to be wise, but at the time there were few clues to warn him of the changes that were looming. Though Africa's first national parks had been declared decades before,[64] the businesslike British had shown little interest in establishing them in Uganda. It would be in the 1950s, another 20 years, before its first national parks were declared. Nonetheless, support for the idea of protecting wildlife was growing, and interest from that perspective, inadvertently spurred on by the king's request for game guards, was gathering around Mburo.

Progress towards regulated conservation was neither steady nor straight, and was resisted by many of the colonial officers who were expected to implement it. Some disliked the limits on their freedom to decide what land was best used for. Others were opposed to what they saw as infringements on the interests of ordinary Africans. And some were still fantastically keen – fanatically keen, it could be said – on hunting, and equally keen to prevent Africans from doing the same. Not surprisingly the first areas established in Uganda for wildlife were 'controlled hunting areas' and 'game reserves', in which 'proper'

---

64 Africa's first national park, the *Parc National des Virunga*, was created in 1925 in eastern Belgium Congo. The park lies along a large part of Uganda's western border with what is now the Democratic Republic of Congo. Thirty years earlier, Hluhluwe Game Reserve was declared in South Africa and was Africa's first formal, government-declared protected area for wildlife.

hunting was sanctioned.[65] In an exercise in control and oppression not unlike that imposed by the Norman conquerors on Saxon England 1,000 years before, the Africans were not allowed to hunt in these areas. In time, though, these reserves, initially set aside for the prerogative of hunting by white settlers and administrators, would fall under more democratic rules that banned hunting for all.

For no obvious reason, perhaps simply due to the perspectives of individual administrators, and in contrast to the growing clamour to protect wildlife elsewhere in Africa, there seems to have been a very pragmatic view of wildlife in Uganda. The first wildlife management body, established in 1923, was the Elephant Control Department, formed to prevent damage to farms. Its officers shot four times more elephants than did the licensed hunters. Though hunting legally required a licence, subsistence hunting was not really controlled, at least not at first. The first step to the active control of what went on in Mburo, other than the historical dictates of the king and the interests that promoted long-horned cattle thus began with the establishment of the Lake Mburo Controlled Hunting Area in 1933, as this is what the king's call for game rangers initiated. It would be years, though, before Mburo was legally gazetted and hunting actively prevented.

A 'controlled hunting area' sounds as though it was somehow intended to protect wildlife. Actually, these areas were established to achieve the opposite. It was only hunting that was controlled in these areas, not the use of the land, and it was understood that once wildlife had been shot out the land would become productive farms or ranches. Those railway cars still needed to be filled, so trade remained the primary agenda and farming the main solution. The sale of hunting licences could be lucrative, but wildlife, especially elephants, was hampering the development of farming. Hunting, reducing damage to farms whilst generating revenue, must certainly have seemed like a good plan.

The control consisted of banning hunting by Africans, which delivered nothing to government, in favour of big game hunting. The government sold licences, the hunters sold meat, hides and ivory, sport hunters paid good money to be shepherded towards their trophy kills, and once the wildlife numbers fell the farms would flourish. Things did not work out quite as planned, though; as the trophy hunters ignored the bushpigs and the baboons that were the main causes of crop damage, farming still failed to take off. At the same time, concerns for the survival of Africa's wildlife were growing in Europe and America. Though it was white settlers and professional hunters that were actually causing the damage, it was hunting by Africans that was labelled the main threat, perversely because Africans hunted for food rather than sport.

It is worth noting that when Europeans started moving around in the interior, they were for the most part seeing Africa after the rinderpest epidemic had swept the

65 As discussed in Chapter 3, the imperialist notion of The Hunt starkly differentiated between hunting by a certain class of white man and hunting by everyone else. The Hunt sanctioned and indeed glorified *sporting* hunting: hunting for pleasure, stalking on foot and delivering a clean kill with a single rifle shot; the prey deserved this, and honour demanded it. Other kinds of hunting were deplored, especially hunting by Africans for food, and were criminalized by the strict licensing of hunting.

continent. Though the virus had destroyed most of the wildlife as well as livestock, the famines, political instability and conflicts that followed allowed the wild animals to recover more quickly than the livestock. The great herds of wild animals that the travellers and colonialists reported encountering during the first decades of the 20th century, and which they proceeded to slaughter with abandon before determining to protect them with equal enthusiasm, had not actually existed before the epidemic; until that point, the customary hunting and grazing had kept wildlife numbers at levels deemed manageable by the African custodians of the land.

This was the case in Mburo too. Nonetheless, lacking this historical perspective and in the spirit of the cultural supremacy of the times, pressure mounted on the British administrators to protect wildlife from the ever-hungry Africans, though this time the immoderate killing of the white hunters was also to be controlled. Given the popularity of hunting amongst the colonial élite and its economic importance for colonial governments, it is surprising that the conservation agenda was taken up so energetically; but so it was, and Uganda too came under pressure to establish national parks.

In 1930 an assessment was made of the potential areas for wildlife conservation in East Africa, in preparation for an international conference to be held in London in 1933. The event was convened by well-connected conservationists, and it brought together senior colonial officers, who were invited to agree to the strict protection of wildlife in their territories.[66] Reserves were to be established, and although differences over how they should be managed were allowed, they were to be exclusionary and permanent. The London Convention, as it is known, in effect bound its signatories to the principles of the national park ideal. Uganda's officials held out for another 25 years before declaring the first national parks, but the idea behind them could not be completely resisted and the principles of exclusive conservation made ground steadily, even in the face of Uganda's status as a protectorate rather than a colony.[67]

The value of areas like Mburo began to be viewed in terms of wildlife rather than farming potential, which then began to be restricted. Mburo had herds of buffalo, zebra, eland and impala. Its lakes were full of hippos and crocodiles. It was famous for its lions – infamous, indeed, as there were not infrequent reports of man-eaters. Soon it was being described as a wildlife paradise to rival any in Africa, and colonial officers were beginning to refer to it in their reports. On top of this, Mburo was sparsely populated. Farming had not really taken hold, and the herders had again fled the area, this time to escape sleeping sickness, another disease carried by tsetse flies – and fatal to humans. Sleeping sickness

---

66 The conference was presided over by the president of the Society for the Preservation of the Wild Fauna of the Empire, recognised as the first international conservation NGO and existing today as Fauna & Flora International. The conference led to the one of the first conservation agreements for Africa, the Convention Relative to the Preservation of Fauna and Flora in the Natural State.

67 Uganda, as a protectorate, had a considerable degree of autonomy over its internal affairs, at least when compared to British colonies. The different kingdoms were, in theory, under the authority of their established governments.

had been so serious in Uganda at the beginning of the 20th century that colonial officials cleared large areas of people for their own protection.[68] The Bahima, meanwhile, left Mburo of their own accord, but government determined that the tsetse fly had nonetheless to be eradicated from that area.

It is not clear why it was that decision was taken. Perhaps plans for the grand ranching scheme, which would start in the 1960s, were already in the offing. Whatever the reason, the attempts to eradicate the tsetse would have lasting impacts on the area. Efforts started in the 1950s and continued for a decade. The first attempt was based on exterminating the wildlife. Tsetse flies are blood-sucking insects, so without their hosts, it was reasoned, they would surely die out. Ignoring the growing clamour to protect wildlife, teams of shooters began to kill just about anything that ran on four legs, large and small. But despite working for years and shooting almost everything, the hunters failed to kill it all.[69] The animals became shyer and more wary in the face of the massacre, and it turned out to be effectively impossible to track down the few that remained.[70] The mighty buffalo and eland might have nowhere to hide, but smaller creatures like bushbuck, reedbuck and bush duiker found places to conceal themselves, as did the nocturnal bushpigs and the ever present but generally invisible leopard.[71]

This should have come as no surprise. Even in areas long settled and farmed, traditional hunters continued to hunt with dogs and nets, and it was clear that however often the wild animals were hunted, some managed to survive – enough, at any rate, that it was worth going out to hunt them.

So it was in Mburo. The hunters were eventually called off, and as the wildlife populations grew back the tsetse fly re-emerged.

---

68 Between 1900 and 1920, Uganda experienced one of the worst recorded epidemics of sleeping sickness, which killed an estimated 200,000 people, mostly living close to Lake Victoria. The Protectorate Government responded with programmes to relocate communities from affected areas. In the 19th century, the Sese Islands of Lake Victoria were densely settled – it was Sese Islanders who had threatened Stanley during his circumnavigation of the lake, chasing him in their war canoes. British administrators cleared the islands, which over time became heavily forested. Outbreaks of sleeping sickness also occurred in areas of western Uganda and also led to clearances, allowing for the subsequent establishment of Queen Elizabeth and Murchison Falls national parks and of game reserves including Mburo.

69 Though the eradication of wildlife failed, it is probable that the programme was responsible for the local extinction of giant forest hog and African wild dog. Both were recorded in the area before but not after the shooting campaign.

70 In England in 2013 the government controversially decided to test the shooting of badgers to control bovine tuberculosis. The target of the programme was to reduce badgers by 70 per cent in relatively small parts of the UK. Even with high levels of investment, infrastructure and support, the cull failed to meet its target, even after the shooting period was extended. The Environment Secretary responsible for the programme famously reported that the marksmen had failed to achieve the required level of killing because 'the badgers moved the goalposts'.

71 Some animals are incredibly sensitive to human presence. Jaguars, for example, avoid areas for weeks or months if they sense the presence of humans. Leopards, however, adapt quickly to human presence and are so good at hiding, they live happily surrounded by people. Stories of leopards living in Kenya's Nairobi suburbs and even down town are common. David Attenborough's Planet Earth II documentary on wildlife in cities included footage of leopards hunting wild pigs in Indian cities. In the 1980s, wildlife counts in South Africa's protected areas were carried out on foot, with hundreds of recorders marching in long lines through the bush. I joined a foot count in Hluhluwe Wildlife Reserve and was told that leopard, though common, had never been counted using this method.

Undismayed, the authorities turned their attention to the bush itself. Tsetse flies breed in bushy areas, and Mburo provided plenty of habitat for them. *Acacia hockii*, for example, can create dense stands of hundreds of young trees growing closely together to form a mass of tangled branches, then as they mature they thin out to form woodland. Termite mounds raise thickets of creepers, bushes and small trees high enough to survive the frequent grass fires. Dwarf forest fills gullies and blankets sheltered hillsides.

By intensive grazing, burning pastures and chopping bush for fences and enclosures, the pastoralists had kept the bush down and the tsetse at bay until they left the area. But the technocrats took an altogether more extreme approach. Having failed to eradicate the fly by removing their animal hosts, it was decided to remove their habitat. Teams of tractors dragging heavy chains brought down trees, large and small, smashing and uprooting the bush, which was then burnt. The destruction was complete; mission accomplished, it seemed, but not for long. The snowy ashes of the burnt trees, washed into the soil by the rains, provided a nitrogen fix that stimulated a burst of growth. The fires had hardly cooled before plants growing energetically from seeds in the soil and re-sprouting from mangled stumps quickly covered the ground. The tsetse flies found this cover quite satisfactory and continued to breed.

The third and final solution was chemical warfare, and Mburo was sprayed with DDT.[72] The tsetse fly was vanquished, for a while at least, but the cost to other forms of life was severe. Every kind of insect perished: butterflies and damselflies, bees, beetles and wasps and the myriad lake flies, along with spiders and scorpions and all the creeping creatures. Without food, the insect-eating birds and mammals died too. Flycatchers that swoop for their prey or pick them from nooks and crevices, woodpeckers, pinion-winged bee-eaters in their rich cinnamon and green colours – these all but disappeared for a time, along with bats, mice, reptiles and amphibians.

Though the losses were quickly made good by animals dispersing into the area from outside, this kind of carnage, born of a mindset that implicitly believed people should and could hold sway over nature and manipulate it to meet their demands, was not unusual then. To be fair, it is not unusual today either. DDT was used across the world well into the 1970s and nature continues to be consumed easily and almost without question to achieve the objectives of individual enrichment or national development. Though such blatant attacks on nature as occurred in Mburo might meet with protests today, similar acts of destruction, though less visible, are still taking place in most parts of the world.

Mburo and its wildlife were treated as a mere platform for development, but the effort to eradicate the tsetse fly was driven, in part at least, by the wish to protect people from the fearful scourge of sleeping sickness. Government simply used the crude tools it had at

---

72 DDT – dichloro-diphenyl-trichloroethane – is a powerful insecticide with long-term environmental impacts due to its persistence in natural systems, resulting in it accumulating in higher levels of the food chain in the tissues of large predators.

its disposal. What justifications can we use today as insect populations are decimated and bee populations crash from our overweening use of pesticides?[73]

With the eradication of the tsetse fly thus achieved, the stage was set for the next round of conflict. In the 1960s Bahima began to return to Mburo, the heart of the Beautiful Land and still the centre of their pastoral ideal. They had fled nagana and sleeping sickness, but now Mburo could once again be the place in which to demonstrate the pinnacle of pastoral achievement that was still their desire. With their return, the sedate red-brown beasts cruised the valleys once more, filling them with the soft sounds of horns gently knocking, of wooden bells clonking around the necks of prize animals, and with the herders' cries of pleasure and songs contentment.[74]

Despite their return, however, their absence had given time for an alternative vision for Mburo to take shape. New actors had arisen with new expectations, and they were ready to impose their interests on the pastoral landscape. The newly independent national government, supported by local élites, agreed that Mburo and its wildlife should be protected. Demands to protect wildlife and wilderness might have started in London 30 years before, but they found support amongst the cadre of young professionals now filling the institutions of independent Uganda.

Wildlife was no longer considered by these actors as vermin that destroyed the crops and livestock of hard-working farmers and ranchers. Nor was wildlife to be the legitimate target of hunters, white or black, in search of a trophy or a meal. A window had opened, if briefly, through which wild animals and their wild habitats were regarded as beautiful. Not as part of the Beautiful Land and its historical and cultural meanings, but as the opposite; Mburo was now to be valued as a 'wilderness', valued for the very absence of the imprint of the people who had formed it and lived in it for centuries. Mburo had attained recognition for what was considered its inherent and essential worth. No longer the Bahima's landscape of pastoral perfection or the colonial landscape of production, Mburo had acquired new meaning in its apparent wildness.

The concept of Mburo as a wilderness, however surprising that might seem given its centuries of use by humans and its cultural significance, demonstrated the triumph of an idea long established at the centre of modern conservation: the imperative of separating wildlife from people. When finally imposed on Mburo, it presented real difficulties to

---

73 Corporations such as Monsanto and Syngenta who profit from neonicotinoid pesticides contest the evidence of negative impacts on bees, just as the makers of DDT denied its negative impacts or argued that the benefits outweighed the damage. The results of a definitive study are awaited, but many believe there is sufficient evidence already and that the precautionary principle should be invoked to ban their use immediately.

74 A man who is able to accumulate a herd of 100 animals is entitled to hang a bell on a prize animal to mark his excellence as a herder and breeder of cattle.

those responsible for implementing it. Love of bush and wildlife might motivate some to conserve nature, but as empire gave way to independence in 1962, the colonial officers and their African counterparts wondered how long this concept could persist. Unless conservation could be explained and justified in ways that were relevant to the newly voting public, they feared that the wilds would be swept aside along with colonial rule.

Arguments that explained and justified conservation in terms of wealth creation seemed the obvious solution. Only this, the bottom line, would, so the thinking went, persuade the local leaders and their people to protect wildlife and wilderness.

This demonstrated an astounding lack of understanding or perhaps interest in how ordinary Africans perceived their natural world, but it accorded well with the colonials' growing preoccupation of the developed nations with economic development and global trade. It also fitted with their assumptions that, in contrast to their own quite recently developed interest in nature's intrinsic values, Africans had always had a purely utilitarian relationship with nature. After all, compared to the elevated sensibilities of Europeans towards the natural world, Africans simply used it for food. Unless wildlife was seen to be essential to economic development, the beasts of the bush would be slaughtered – and slaughtered, to boot, in horrible, primitive ways in unsustainable numbers.

The solution those colonials arrived at was not to inculcate their own admittedly rather confused and conflicted affections for nature into the African mind, but to construct arguments for conservation around wildlife's economic value. No doubt they were right – even if for the wrong reasons – to avoid making efforts to graft foreign values onto post-colonial Africa; but rather than investigate how people actually related to nature, they reconstructed and rebranded the African wilderness and its wildlife as a natural resource to be tapped. The conversion of Mburo into a game reserve was set in motion in 1963, one year after independence – and the returning Bahima found they were unwelcome in their own land.

Though the strengthening of the control over Mburo fitted the resource-based direction that conservation was heading in, the officials responsible for Uganda's wildlife would take things in a different direction. Uganda's wildlife professionals, attaching themselves to the ideas being promulgated by international conservation organizations helped them carve out domains of influence that eased them through the uncertainties and opportunities of independence. The ideal of the exclusive protected area might have little to do with traditional African perspectives, and it would not promote sustainable resource use, but it was a powerful tool for securing control over land, and authority over people.

So the battle lines between Bahima and conservationists were set, and Mburo was to become their battleground. The demands for exclusion might have been based on foreign ideas, but the Ugandan government took them up and wielded them vigorously, and in 1963 Lake Mburo Game Reserve was formally created. Its 700 square kilometres was small by the standards of the day, but it included the most praised, the sweetest, the most desired grazing lands of Ankole. Lesser men might have picked a less contentious place to test their strength and the nation's commitment to conservation.

The history of antagonism between the Bahima and the protectorate government over the game reserve had sown the seeds of conflict that would escalate with the creation of the national park 20 years later. For the Bahima, the pastoral lands of Mburo existed entirely for the purpose of grazing long-horned cattle; it was this purpose – and, as importantly, the absence of any other purpose – that gave meaning to the land. While the Bahima depended on their cattle for life itself, they bred them to honour their Bachwezi ancestors.

In contrast, the governing interest of the British in production and trade had opened Mburo to farmers and ranchers. By the time it had begun to be regarded as a wildlife paradise, it was already scattered with farms, fishing villages and ranches, as well as with traders, shopkeepers, bicycle repairmen and all the bustle of rural life. The gates that had been kept firmly shut for centuries by Ankole's pastoral tradition had been forced open by the elevation of production to an imperative, combined with the relegation of the Bahima's pastoral ideal to a primitive and ignorant obstinacy.

Escalation of exclusion

The 1960s and 1970s had seen an escalation of the conflict between the Bahima and officials. The Bahima had not been powerful enough to force the Game Department out, especially after the loss of their king,[75] but they could impose themselves and their cattle on the land and, with increasing confidence and in increasing numbers, this is what they did. Though a coolness existed between the educated Bahima and those who had retained the nomadic lifestyle, they had enough in common to work together; although the élite held positions of power in local and national government they retained an attachment to the long-horned cattle and to the land, if not with quite the same exclusive devotion as the traditionalists.

As great swathes of land were enclosed as ranches and farms for the élite, many argued that Mburo should follow. The reserve would certainly have looked like a prize for the picking, and within a decade a quarter of the game reserve had been handed out as ranches to individuals and institutions. Learning that possession was nine tenths of the law, some pastoral families began to fence off land in the reserve and even to cultivate parts of it.[76]

Despite the protests of the Game Department, neither the incursions of cattle nor the expansion of farms stopped. Aiming at what it thought would be the softer target,

---

75 When independent Uganda waved goodbye to the British the country inherited a complex political system of government modelled in part on Britain's system of constitutional monarchy and parliamentary democracy. Milton Obote, Uganda's first prime minister, accepted a constitution that gave a degree of independence to four traditional kingdoms. Buganda was the most powerful of these and the Kabaka, Muteesa II, became Uganda's first non-executive president. This unwieldy solution did not last long, and by 1966 Obote had forced through a new constitution and abolished the hereditary kingships. This was a grave blow not only to democracy in Uganda but also to many cultural institutions that were closely related to the political and spiritual powers of the kings.

76 Early adopters of farming amongst the Bahima made use of paid workers rather than actually undertaking cultivation themselves. Until quite recently, the jobs that family members were responsible for remained largely unchanged, despite the growth of farming: men undertook all tasks associated with caring for the cattle, while women were responsible for the milk and for caring for the home.

the Game Department focused its efforts on removing the pastoralists, and in the years immediately before and after independence the movement of pastoralists was increasingly restricted. Previously the herds had been allowed to journey through game reserves, but now what was termed 'camping' in those areas was banned. The quaint notion of the Bahima camping suggests a rather obscure conception of their lives.[77]

Unlike transhumant pastoralists, who make annual migrations between dry and rainy season pastures, the Bahima had been truly nomadic. Their movements, under the control of elders, were determined by the vagaries of rainfall and grazing, avoidance of tick and tsetse infestations, and by omens and auguries. The whole land was open to them, and they moved as they wished in family groups. An eka might be home for months or even years depending on conditions, but on the move or grazing far from home, the men would bivouac with their cattle. The prohibition against 'camping' in the parks and reserves was probably not intended to undermine that way of life but to make it clear that those areas were out of bounds.

As Bahima started spending more time in Mburo, confrontations between the spear-carrying herdsmen and the gun-carrying game rangers took place with increasing frequency. Cattle were driven into the reserve at night from adjoining[78] lands, and families established eka in remote corners of the reserve. Bahima were threatened, and threatened back. Cattle were impounded, fights broke out, and injuries were sustained on both sides.

The struggle over Mburo was not just over land and resources. It was a struggle between the old ways and the new, between the kingdom of Nkore and the nation of Uganda, between traditional values and the ideals of a westernized élite. The contest over Mburo was also a fight between two newly powerful groups, the officers of the Game Department pitting their strength against the Bahima élite, competing for power in the emerging institutional landscape of the Ugandan republic. For the officers, victory in this contest would give them control over land, water, grazing and wildlife, as well as secure government budgets and political power.

Taken directly from globally recognized practice, the exclusive protected areas helped the Game Department strengthen its claim as the sole controller of large areas of land. Their resistance to the reserve became a weapon that the pastoral Bahima wielded to protect their way of life and values and which the élite Bahima used to secure their private estates.

The farmers in Mburo were actually a greater threat to the wildlife landscape than were the pastoralists, but actions against the farmers were less severe; indeed, while the

---

77 Following the pattern of many of the Bantu group of languages, in which *Bantu* means people while *Muntu* means person, *Bahima* is plural or refers to the Bahima people, while *Muhima* is singular, a *Muhima* man or woman. Similarly, *Batwa*, the forest people of Uganda, is plural while *Mutwa* is singular.

78 Complex interrelationships between families were signified by the exchange of cattle. This meant that many of the cattle in Mburo belonged to other families. Exploiting this system to take advantage of opportunities in Mburo, wealthy Bahima would 'give' whole herds to friends or workers to graze. This made it difficult to determine whose animals were whose, greatly adding to the difficulties of enforcing the law excluding cattle from the reserve and later, the national park.

steps to control the pastoralist incursions intensified, the farmers were issued with permits to remain. After all, it had not been that long since some, at least, had been invited to settle. But these permits would cause considerable difficulties for the Game Department and the national park in the future, as they failed to mention how much land could be cultivated and said nothing about the status of future generations. Not surprisingly, farmers interpreted their permits generously. If they wanted more land they cleared more; when their children needed land, they gave it to them; and if friends or relatives asked to join them, families happily agreed and gave them land too.[79]

When I first flew over Mburo in 1981, bumped and buffeted in that small plane, I could see that the Beautiful Land was already compromised, no longer the fenceless, farm-less expanse of the Bahima's imagination. Much open land remained, however, and a delicate network of paths linked eka and farms to the small trading centres strung along the few tracks that meandered from the main road to the lake. Along them ran pickups and bicycles carrying fish from the landing sites or goods to the farming settlements. Where the soils allowed, the gentle slopes at the heads and along the flanks of valleys had been planted, and banana plantations appeared as emerald swatches within the olive drab of grasses and woodlands; bunches of bananas flowed to markets in the growing towns.

At that time a new breed of farmer cattle-keepers was challenging the ancient separation between pastoralists and farmers. The Game Department feared that without a

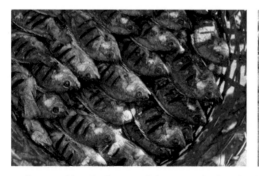 

Left: Dried haplochromis fish are packed in baskets

Right: Large bunches of matoke cooking bananas are transported from farm to market by bike

---

79 There were good reasons for farming communities to want to grow. Small or isolated farmers were more vulnerable to damage by wildlife. The larger the settlements, the less the risk to individual families, especially to the families located in the centre of growing communities, surrounded by a protective buffer of more newly established families.

strengthened resolve, Mburo and its wildlife would be lost. Its proposal to gazette a national park had found favour with the district government and conservation organizations, and by the time I flew over Mburo the discussions were already well advanced. I looked down on the rippling lakes and grassy hills and valleys with the cattle mingling with the wildlife, the contained communities of farmers and fishermen, and thought 'this looks right'. But at that very moment the plans to end it were being developed – plans that would soon impose the separation between people and nature that is demanded by national park status in most parts of the world.[80]

---

80 The areas classed as 'national parks' in Britain are unusual in being widely farmed and settled. Though they do not follow the general rule of excluding people from living within them, they are increasingly criticised for their low levels of biodiversity. The over-grazing of these lands, especially by sheep and non-native deer, is widely considered the chief culprit in their failure to protect nature.

3
Parks for the world

My first flight over in Mburo in 1981 left me entranced and excited by a glimpse of a landscape in which production, consumption and nature seemed in harmony. I saw people in their modest homes, children rushing out to watch and wave as we flew low over them; there were herds of cattle scattered through the bush, looking as though modelled from the clay of the termite hills they rubbed against; and there was wildlife – herds of zebra and impala scattering under us, hippos plunging away in the shallows of the lakes. Strangely, given that I was 8,000 kilometres from home, it did not seem strange to me at all. It did not seem so different from my home in the Weald, where the land was also used and cherished at the same time, full of people pursuing their lives and needs of all sorts. Why should not Mburo continue like that?

At the time I probably thought my imaginings signified nothing more than my ignorance of practical conservation. Certainly, I knew nothing of the history of Mburo, the fears of officials over the loss of Mburo's wildlife and the erosion of its earlier untrammelled and expansive character. I was learning that, in Uganda at least, mixing people and nature was not considered possible, or even a good thing in principle. But even then, just a few months old in Africa, I felt there was something wrong and that we conservationists – I considered myself part of that group already – were failing to recognize something that could help us achieve our objectives; that in fact people and wildlife *could* mix. But it was too late, for Mburo at least. The ideal of exclusive conservation landscapes had since the colonial era been firmly established in Uganda. The model had been drawn into the institutions of the independent nation, and was about to be fiercely and relentlessly imposed on Mburo and its people.[81]

I was not present in Mburo in 1983, but can envisage what happened. Imagine heavy banks of cloud in shades of purple and grey as they slide across each other. Snatches of blue

---

81 The ideal of the strictly protected area that excludes most human activities and presence seems to exert a strong attraction for conservation policy makers. The European Union's ten-year Biodiversity Strategy published in 2020 focuses on 'establishing a larger EU-wide network of protected areas on land and at sea, building upon existing Natura 2000 areas, with strict protection for areas of very high biodiversity and climate value'. 'Strict protection' means, once again, the exclusion of people.

sky are open in the distance where the sun still shines. A small plane is flying somewhere overhead; the warden is directing operations from above by radio. The storm approaches and a soggy, dirty-white blanket falls across the land. Rangers arrive at a collection of huts, moving slowly under the trees and through the compounds. A thin mist flattens the shapes of the hills and hides even the storm itself, purging depth and perspective. Low thunder rumbles like a trolley running over concrete.

Some villagers emerge from their houses, possessions tied in bundles with sisal cords and bush rope. Most of the houses are already empty. Few have waited for the rangers to come. They know how these operations go. They cannot protect their homes, and prefer not to watch. Flames begin to curl from the windows and doors left open by the departing owners. The villagers are hurried towards a truck parked somewhere nearby. The blaze licks the eaves and punches through the grass thatch, which blackens and collapses down through the roof poles. The first heavy gobbets of rain smack the ground, turning quickly into an intense, noisy downpour. The light dies, drawn from the land, and the day turns gritty and dark. The rain, falling heavily, merges into sheets of water that move across the ground, first in runnels, then in torrents of cinnamon like a horrible soup, spilled and gushing to fill every depression.

When the rain stops and the rangers have moved on, villages established 20 or 30 years before – some earlier, solid, substantial houses of mud and thatch, fruit trees and banana plantations – are in ruins. The mounds and hummocks will soon be lost under weeds and scrambling bushes; termites will eat up the building poles that escaped the fires, and where the houses once stood is marked only by dark-leaved mango trees.

The decision to create Lake Mburo National Park had been taken in 1982, before I left Uganda. As I then travelled south through Africa, passing through Rwanda and Burundi and via Lake Tanganyika into Zaire, down to Zambia, into Botswana and eventually to South Africa, finding my way and my feet in this vast, exciting and challenging continent, Lake Mburo, and indeed Uganda itself, were slipping into chaos. The creation of the park would set the authorities firmly against the people, igniting the fiercest battle yet for the control of Mburo and its resources, and for its values and meanings.

Fifteen years later, in 1997, I sifted through government archives, leafing through files of yellowing letters and reports, some half-consumed by termites, to learn how the decision had come about. The meetings, recorded in detailed minutes, provided a clear picture.

Suited politicians and civil servants gather, conversing quietly in the cool, high-ceilinged chambers of Mbarara Town's municipal offices. Light floods in shafts through small windows and wooden shutters to fall on worn desks and polished red floors, illuminating beams of dust particles. Solitary mud-dauber wasps levitate silently to their nests on window frames and roof joists, yellow legs dangling as they drift upwards.[82] Soft voices bemoan the decline of Mburo's wildlife and berate the lawless pastoralists who roam the area with impunity. Peasants with nets and spears are harrying the impala. The law and its officers, especially those of the Game Department, are becoming a joke. The pressure from Kampala is gentle but insistent.

Agreement to uplift the reserve to a national park does not take long. There are no dissenting voices; why, indeed, would there be? Uganda already has three celebrated parks created by the British; national and international experts have documented Mburo's importance for wildlife, and for years have been arguing for stronger protection. The reserve is being overrun by farmers, fishermen and cattle-keepers. It is only the creation of a national park there that will focus attention on the problem, build the resolve necessary to deal with the problem, and provide the legislative strength needed to succeed.

Once the decision is made, actions follow fast. From this point, the removal of the people from Mburo is inevitable. The law requires it, normal international practice accepts it, and conservation success apparently demands it. The expectations of a national park are unambiguous, and the establishment of the park will naturally follow the norms of the conservation ethos that separates local people and nature to create 'wilderness'. The government's political agenda to punish the pastoralists is not discussed.

That decisions leading to arrangements so unfair and unaccountable could be made in the name of conservation, that acts of such harshness could be carried out to protect impalas, seems extraordinary by wider standards. Protecting nature is after all a laudable pursuit, carried out for well-established reasons for the good of humanity and nature. Though voices are raised against such acts both within and without the conservation movement, then and even now, it seems that demands to protect nature, wildlife and biological diversity overrule other considerations, even such damage to communities.[83] Omelettes, I hear dispassionate voices explain, cannot be made without the breaking of eggs, or even heads.

What happened at Mburo was a direct and predetermined consequence of declaring the national park; a cascade of actions inevitably followed, the law stepping in to demand and direct the course of these actions. The law in question was the Uganda National Parks Act, promulgated just two years after independence. Heavily influenced by British conservationists, that law was also based on guidance from the International Union for the Conservation of Nature (IUCN). The union, established in 1948, guides and supports

---

82 Though the decision to gazette the national park was taken by Parliament, the process began in the Game Department before being taken up by the Ministry of Tourism and Wildlife. This was before districts became decentralized authorities but after the days of the Kingdoms. Nonetheless, District administrations would have been consulted on decisions with significant local impacts.

83 See, as an example, the July 2018 article by Souparna Lahiri, 'Saving tigers, killing people', to see that terrible actions are still taken in the name of conservation. https://www.aljazeera.com/indepth/opinion/saving-tigers-killing-people-180703110004941.html

the delivery of nature conservation. It has played a central role in defining protected areas and encouraging governments to adopt them to achieve conservation targets. Today, seven categories of protected area are listed, and governments are expected to reflect these in their national legislation. At one end of the scale are the strict and exclusionary Nature Reserves, Wilderness Areas and National Parks, which prevent almost all forms of use of land and resources. At the other end, sustainable resource use is encouraged, and human habitation allowed for.

When Mburo was established, these more accessible forms of protected area had not been defined, and the legal options were more restricted than those of today. But the decision to create a national park was taken with the express intention of halting the farming, fishing, and grazing that was going on, and outlawing the villages that were there. So even if less restrictive categories had been available, it seems unlikely that they would have been chosen. If the suggested political motivations were true, the national park, by wresting control of Mburo from the Bahima, was also the weapon of choice for attacking them.

It is easy in retrospect to point out that other forms of conservation area could have been employed to protect Mburo. Knowledge of how indigenous and local communities' beliefs and practices can protect nature is not new, but it has taken decades for their potential contributions to be recognised. The Indigenous and Community Conserved Area Consortium[84] has led the way in raising the profile of areas protected by communities for their own purposes. Growing awareness of these has led to the recent inclusion of the somewhat oddly named 'Other Effective Area-Based Conservation Measures' in processes to deliver the latest global target for protecting the planet.[85]

At the time when Mburo was gazetted in 1983, though, governments were almost the sole players in delivering conservation, and they were largely unchallenged by civil society. The initiation of a conservation action was perceived as primarily a legal process, and the legal possibilities were limited.

Today, however, the situation is vastly different, and perspectives on protected areas have broadened dramatically. National and international law remains, however, at their centre. But laws give us only one perspective of what conservation is and what protected areas are for – and quite a narrow perspective at that. Once the interests of local communities are inserted into the discussion, the definitions of what the protected areas are for and how they can be managed must be opened to positions and perspectives that go well beyond those of governments and conservationists.

---

84 The ICCA Consortium was established to recognise the importance of indigenous and community-controlled areas for conservation, to champion the communities responsible for them, and to help prevent them begin subsumed or co-opted within government programmes or eroded by social development and unsustainable use. https://www.iccaconsortium.org

85 Reference to 'other effective area-based conservation measures' first appeared in the text of the Convention on Biodiversity's Aichi Target, which aims to increase the global cover of protected areas through 'well-connected systems of protected areas and other effective area-based conservation measures'. Considerable efforts have been invested subsequently to understand what these might actually be. https://www.iucn.org/theme/protected-areas/wcpa/what-we-do/other-effective-area-based-conservation-measures-oecms

Left: A herd of Ankole cows, moving though the valleys of Nshaara as the King's herds moved during the times of the Nkore kingdom

Right: Ankole calves and zebras happy in close proximity

The days when Mburo had been infused with the intensity of pure Bahima pastoralism, when the area might have been considered a community-conserved area, were long gone. The king's magnificent herds no longer cruised uncontested the seas of waving grasses accompanied by flotillas of snowy white cattle egrets. By 1983 the lands around the lake supported a diverse set of people busy with many activities, and had done so for half a century.

The mixture of land uses certainly resulted in competition and conflict, but not to such a degree that the atmosphere of peaceful coexistence between the natural and human worlds was entirely lost. The impala might move aside as methodically grazing cattle pushed through them, but then they would close in behind to feed on tender shoots exposed by the cattle's tough mouths rending the coarse grasses. Although bushpigs raided crops, farmers working together could trap or shoot them, getting meat into the bargain. The pastoralists might complain that farms disfigured the land and encroached on rich pastures, and the farmers might complain that the cattle trampled their crops or damaged their fences, but by and large all had learned to cooperate.

But all this would suddenly and dramatically change, and the slow evolution of this mixing of peoples and activities and land uses would be summarily thrown into reverse. The calls of the boys herding the cattle and the recitations of the elders praising the most beautiful were replaced by the shouts of rangers, rifle shots and the hurrying of hooves as the herds were driven from the park or into pens, to be ransomed later. The quiet chatter of fishermen lifting their nets in the early morning gave way to cries of distress as their smoking ovens were thrown down and their nets thrown into the fires. The clearance of the park had begun.

Proclamations were issued ordering everyone living in the new park to leave. There was no discussion, no debating the fairness of the decision, no questioning of the law. The decision to make Mburo a national park had been taken, and in it there was now no place for people, their farms or their animals. Nobody troubled to enquire too closely how all these people had found their way into the reserve, whether they had broken any laws by settling there, or whether they were there legitimately. Donald Mugume, a neighbour – and, surprisingly, a supporter of the park – was one of the many evicted to make way for the park despite his pleas. and one of those who reclaimed his land when the national park fell in 1985. Today he lives near the current boundary, raising his cattle and growing his bananas.

'My great-great-grandfather was given our land as a reward for supporting the king in his struggle to defeat a brother to retain the Ankole drum of kingship,' he explained to me. 'This was before the British came. In time, as things changed under the British, a land title was issued to us. The land from the king was now our land – for ever, we thought. Later, my family ran from the land to save our cattle from nagana and the tsetse fly. My grandfather lived the life of a nomad, but he went, with his papers, and when the threat of disease was gone he returned to our land. But then, when the park was made, our title counted for nothing. We were outside the law of the park, and had to leave. They called us "squatters" on our own land.'[86]

There was anger and resistance; injunctions were sought in courts and petitions submitted to ministers; political connections were worked, and arguments raised against the park; but all without success. The park had been declared legally, and the government would have its park.

Uganda's most seasoned warden was given the job of setting up the park as decreed by Parliament, and he set about it with enthusiasm. Before we condemn him, we must recognize that he was following the standards of the time and the place. He had trained, and served under, the British wildlife officers just a decade before, and was applying what he had learned within the well-established culture of the wildlife sector. Further, he was pursuing the vision of the Uganda National Parks and its directors, formulated and written down as policies, delivered through well-honed strategies and practices, and sanctioned by law. We must not forget, either, that the vision for Mburo as a national park was encouraged, promoted and supported by international conservation organizations.

---

86 The term 'squatter' was and still is used to describe someone who has settled on land owned by someone else. In Uganda, land law is complicated. Freehold land has existed since colonial times but overlays traditional rules of access and use. The line between tenant farmers and squatters is a fine one. At different times, households have moved into protected areas, and government, as the owner of these lands, defines them as squatters. If it demands their removal, it might agree to compensate them for their crops and assets, but no compensation will be paid for the land.

People found living within the borders of the new park were all treated in the same way; rich or poor, powerful or weak – they all lost their land. Whether they were land title holders like Mugume, or farmers who had been invited by British administrators to settle 50 years earlier, or householders issued permits when the game reserve had been gazetted in 1962, or whether they had arrived just a year or a month before to grab a piece of the disintegrating game reserve, all were branded 'squatters'; all were considered to be illegally present, and all were evicted. Shopkeepers in trading centres, ranchers on great ranches, fishermen on their landings, all were given orders to leave, and when the deadline expired their homes were destroyed.

From the day the park was gazetted, the warden and his men worked hard. They built roads and tracks; they established ranger posts and gates at strategic points; they erected radio towers and masts; they constructed a headquarters on the top of Rwonyo Hill looking out across the lake. And they patrolled the boundaries, arresting the hunters and herders and residents who resisted eviction. In short, they undertook the steps deemed necessary to establish and maintain a park, following the pattern set by Uganda's first parks established 30 years earlier, and the lead of Yellowstone National Park 80 years before that.

But time was not on their side. Despite their determined efforts, the park had not even been fully cleared of its residents before the arrival of the National Resistance Army, when it fell under the control of the rebels and their followers in 1985. Some of Museveni's army were amongst those that had been evicted, others had been promised land, and all of them believed that it was wrong to have established the park in the first place. Indeed, the political campaign that underpinned the military campaign of the National Resistance Movement – the rebels had issued a ten-point agenda for their movement – included the injustice of the creation of the national park.

After the battle for Mburo had been entered into, and lost by the conservationists, there was no space for discussion. The opportunity to propose a compromise which would have allowed the interests of conservation and community to stand side by side had passed. A solution modelled on a partnership to reconstruct the kind of mixed land use that I had seen might have worked before, but now that could not even be suggested. The best that the defeated forces of conservation could hope for was to salvage something of the park while, hopefully, learning a few lessons from its rapid rise and fall.

Given the circumstances that prevailed, it is surprising that Mburo was not degazetted completely. The new government listened to arguments that such a move would threaten all of Uganda's national parks as well as its game and forest reserves. If the president was seen to give the park away to his people, a precedent would be set.

A commission was established to investigate the injustices of Mburo's creation and propose a solution that would recognize the rights of the
, meet the expectations of those who claimed land, and respond to the arguments for keeping a park. It was not going to be easy to square this stubborn circle.

The commission began by restoring ranches from the 1960s scheme to their owners. Next, a resettlement scheme was set up, and the families who had been displaced and dispossessed by the war were given land to settle on. And finally land was made available to returning settlers, newcomers and, one way or another, supporters of the new government. When the cuts had been made, little more than a third of the park was left.

Despite this act of Solomon, much remained unresolved that made the effective management of the remaining area nearly impossible. Households that could show they owned land in the area that remained part of the park were told that a solution would be found for them, and that in the meantime they should simply stay where they were and carry on with their lives. This solution was not dissimilar to the decision to issue permits to families when the Game Reserve had been established 20 years before. It had created problems then and would create the same problems again. But there is little doubt that without some sort of an intervention time, attrition and the gradual nibbling away would have ended Mburo as a place of nature conservation.

When I began work in Mburo in 1991, in an effort to help save what remained of the park the warden and I approached many people and heard many accounts of what had transpired during the clearing. We pieced together a picture of events. If the warden and his team were going to overcome the fierce negativity toward the park they would need a better appreciation of what had happened, how people had suffered, and how they had responded. In addition, as we searched for means by which the park could be recovered, we had to make sure we did not repeat the same mistakes. Of course, many of the circumstances of Mburo's story are particular to the place and its people. The Bahima's strong attachment to the land had an especially strong influence on their relations to the authorities, as well as to the idea of conservation. But above and beyond the particulars of time and place, it was the way the conservation agenda had been translated into practice that had played the central role in the débâcle. If I was going to do anything useful at Mburo – indeed, if I were to continue to work in conservation at all – I needed to reconcile my wish to protect nature with the contradictions and conflicts that seemed embedded in the entire venture.

Back when Mburo had been proposed as a national park, few had contested the belief that strictly protected areas were essential to conserve nature in the long run; the practice of reserving land exclusively for nature had been tried and tested for over 100 years, and it was at the centre of the modern conservation movement. Uganda's first two parks had removed their communities from their borders.[87] The third, Kidepo Valley National Park, which had been the hunting grounds of the Ik people, caused well-documented suffering and loss, and resulted in their decline and marginalization as a people.[88]

The British authorities responsible for making those parks were simply following in the spirit of the times and the movement. The first national parks in the United States had been established on the lands of Native Americans who had been banished from them, with soldiers put in place to ensure they did not return. Americans were still being removed to make national parks as late as the 1930s – and not just the powerless Native Americans, but white communities too.[89] So, the steps taken to establish Lake Mburo National Park by clearing it of its people had precedents both within and outside Uganda.

The ideal that demands the eviction of communities had come from the first parks created under the ideology of modern conservation. This movement has by many measures been extraordinarily successful; since the creation of the Yellowstone National Park in 1872 in the USA, the movement has encouraged, cajoled and compelled the creation of similar parks all around the world. Governments legally protect close to a fifth of the surface of the planet, and the area continues to grow.[90] Many consider the global network of parks, reserves and protected landscapes, one of the great achievements of the 20th century, the crown jewels of the conservation movement. But despite the successes of the exclusionary parks, their implications raise questions over their legitimacy. Unless these questions are resolved, contention will hang over the idea of protected areas and may eventually dent the achievements of the conservation movement.

The importance of the national park ideal generally, and to the establishment of Lake Mburo National Park in particular, suggests that we should examine the ideas that

---

87 The warden that cleared Lake Mburo had previously managed Queen Elizabeth National Park. When he discovered two villages of Banyabatumbi people living in the park he immediately evicted them, even though they had lived there for generations. The Banyabatumbi, who lived by trapping fish in rivers and lagoons along lake edges, were left in place when the park was established, but the warden removed them to the edge of the park, where they remained destitute. It required the intervention of local churches, which donated food and shelter, to save them. Eventually pressure from church leaders persuaded the warden to let the Banyabatumbi join a fish-landing community in the park. They became fishermen there, but were not allowed to return to their village sites or resume their traditional way of life.

88 The anthropologist Colin Turnbull wrote about the impacts of the 1962 creation of Kidepo Valley National Park on the Ik people in his controversial book, *The Mountain People*. Turnbull, C.M. (1972) New York: Simon & Schuster.

89 The history of the creation of the first parks, including Yellowstone and Yosemite Valley, records the struggle of certain individuals to go on living in them. When Shenandoah National Park was established in 1935 in the State of Virginia, thousands of farmers were evicted against their will. Their story is movingly told by Sue Eisenfeld in *Shenandoah: A Story of Conservation and Betrayal*, Eisenfeld, S. (2015) University of Nebraska Press, p. 205.

90 The Aichi targets, established under the Convention on Biodiversity in 2010, set targets for 17 per cent of the terrestrial area and 10 per cent of coastal and marine areas to be protected by 2020.

underpin it and many of the practices designed to uphold it. This will help explain the decision made to protect Mburo, and how the actions taken could be justified. The history of the national park ideal may also offer some explanation for how a movement originally inspired by the deep spirituality of connections between people and nature could evolve to require, at almost any cost, their complete separation. This conundrum, which originated with the beginnings of modern conservation and still persists, is responsible for many of the moral and practical problems that plague the conservation movement. Further, it is the foundation on which the practices that have dealt out suffering and loss to so many have been constructed.

The founding fathers of modern conservation, agitating and organizing with remarkable success in 19th-century America, had very different perspectives from those that guide conservation today. Their ideas and beliefs, resulting as they did in the establishment of the first national parks, have over the decades evolved into the ethos of the institution that dominates the modern conservation movement today.

Its current values did not arise in a cultural vacuum, or indeed take form without forerunners. The ideas that inspired the champions of the first parks had deep roots in intellectual and cultural movements in Europe. These crossed the Atlantic to inform the development of values in the newly forming nation.

William Wordsworth, the romantic poet who some people credit with inspiring the conservation movement in England, chose to live a quiet life, removed from the increasingly insistent business of urban life, to achieve the tranquillity and inspiration he needed in order to write. Nature was the space in which he created, as well as being the source of his creativity. The romantic movement, of which Wordsworth was a luminary, gave western culture a new way of thinking about nature. Reacting against the industrial revolution and to scientific modes of thought, Wordsworth believed in a symbiotic and spiritual community between humans and nature, with intuition, empathy and emotion lying at its heart.

Wordsworth and a community of poets, artists and thinkers had a strong influence on American culture. Henry Thoreau, philosopher, naturalist, writer and pillar of America's evolving relationship with nature, was strongly influenced by romanticism. He to sought a retired life, and spent years living on the edge of a remote lake, following a simple existence immersed in nature. 'An early-morning walk is a blessing for the whole day,' he wrote, a succinct expression of his bond with nature and his efforts to live calmly and respectfully within it.

John Muir, the Scotsman celebrated as the originator of the movement to protect wild lands and landscapes and the creation of America's first national parks, was greatly influenced by Thoreau.[91] For Muir and his followers, nature was beauty, nature was sublime, nature was the divine creation. Muir wrote of his longings for nature and the window it opened on his soul. He extolled its energizing and restorative effects, and believed that immersion in it would make people better and stronger, physically, mentally and morally.

'Everybody needs beauty as well as bread, places to play in and pray in, where nature may heal and give strength to body and soul alike' he wrote of his beloved Yosemite Valley.[92]

Religion and a profound spirituality lay at the centre of his love of nature, as it did at Wordsworth's and Thoreau's. This should not surprise, as they lived in a time when religion was central to social, intellectual and political life. So the contrast between the values that inspired those early proponents of nature's protection and those that inspire the current conservation movement seem, at least on the surface, to be enormous.

Some of today's proponents of conservation have been persuaded, it seems, to consider nature as little more than a collection of species in a space interacting with terrain and climate to produce various goods and services that can be bought and sold. This is the language of global finance aligned with much of today's global political analysis. But it is not just the language of conservation that has changed; an entire way of relating to nature through passion, sensation, compassion and emotion seems to have been replaced by one based on the material worth of the benefits that nature brings.[93]

Muir and his supporters demanded protection for places where nature's capacity to inspire awe was most outstanding. As more and more settlers arrived from the eastern regions of the continent, they came to believe that the demands for land and wealth, closely aligned with the national imperative to grow and strengthen, would unless checked destroy these riches. Even when Muir first journeyed to Yosemite, walking inland from San Francisco in 1868, he notes, 'Cattle and cultivation were making few scars as yet', foretelling a future he feared when they would.

---

91 Though John Muir is considered the father of America's national parks, and therefore perhaps of the modern conservation movement, it was George Catlin, a painter, who first put forward the idea. Catlin had lived with and painted Native American tribes and the world they inhabited. In his *Letters and Notes on the Manners, Customs, and Condition of the North American Indians*, he wrote in 1841 'And what a splendid contemplation too, when one (who has travelled these realms, and can duly appreciate them) imagines them as they might in future be seen (by some great protecting policy of government) preserved in their pristine beauty and wildness, in a magnificent park, where the world could see for ages to come, the native Indian in his classic attire, galloping his wild horse, with sinewy bow, and shield and lance, amid the fleeting herds of elks and buffaloes. What a beautiful and thrilling specimen for America to preserve and hold up to the view of her refined citizens and the world, in future ages! A Nations Park, containing man and beast, in all the wild and freshness of their nature's beauty!' Though his vision of preserving the culture of the tribes in his nations park might seem wrong or strange today, retaining the connection between people and nature lay at the heart of his vision. Had his idea been taken up, how differently might the evolution of modern conservation have been?
Letters and Notes on the Manners, Customs, and Condition of the North American Indians, Catlin, G.E.O. (1857) Philadelphia: Willis P. Hazard.

92 *The Yosemite* describes the landscape and Muir's life within it from his first visit in 1868. He saw Nature and the Yosemite as his home, and it was this idea that he developed to lobby for its protection.
The Yosemite, Muir, J. (1912) New York: The Century Company.

93 A basic principle of the Ecosystem Services approach to valuing nature is that there are no benefits from nature unless they are consumed by people.

It was the fear that their sacred places would be lost that led these men to argue for the creation of parks from which people needed to be excluded. Their heartfelt intentions to protect places of beauty and power would be wildly successful but, transposed to sites around the world, they have too often served to separate rather than unite people and nature. The experience of profound connections with nature that inspired and motivated these men would be denied to people across the globe whose connection to nature were also profound, also spiritual, and often lay at the core of their identity.

The spread of national parks and protected areas around the world was carried on the shoulders of three notions – wilderness, countryside and the Hunt. Unrelated to modern ecology or informed by science, these ideas fused to create an ideal that in due course would allow or even demand the actions of the wardens and rangers who established Lake Mburo National Park at the point of a gun. These ideas evolved in different places and at different times but came together to create a mindset in which exclusive protected areas could be established almost without regard for the consequences for people affected by them.

Since Yellowstone National Park was established in 1872 to protect 9,000 square kilometres of forested mountains in the western United States, the idea of wilderness has had a powerful influence on conservation. Wilderness is not an ecosystem or an environment or a habitat with characteristics that can be described and defined, but is defined by a notion of 'naturalness'. The word is almost, though not quite the same as, 'wildness', which implies a place outside human control. Though different interests and organizations may highlight different aspects of the idea of wilderness, conservationists are happy to think of it as not only land without people but as land without evidence or signs that they were ever present.[94] Wilderness is pure, nature untouched, a world unpolluted by humans. But in truth there is nowhere on earth untouched by humanity, except perhaps the frozen poles, the ocean deeps or the mountain peaks.[95]

Transplanted into Africa in the late 19th and early 20th centuries, the idea of wilderness absorbed values related to European ideas about domestic landscapes, especially the particularly English idea of 'countryside', which combined the romantic movement's engagement with nature and humanity with the politics of class and class privilege. The

94 IUCN describes a Wilderness Area, one of its seven categories of protected area, as being largely unmodified, as retaining its natural character and without permanent or significant human habitation, managed to preserve its natural condition. This is as close as IUCN gets to defining wilderness in conservation terms. National Parks are described simply as natural or near-natural areas.

95 Today, not even these are unsullied, and signs of pollution and plastics are found everywhere including at the bottom of the oceans and in the ice cores of Antarctica. Indeed, scientists have determined that the planet has entered a new epoch, the Anthropocene, described by the all-pervasive influence of humans.

third element derives from the colonial experience, the imperial invention of the big game hunt, which merged the rules of class and control to create a whole new idea of supremacy. As colonial hunting underpins much of the history of state intervention and protection of Mburo, it is worth looking at it in some detail.

The untrammelled emptiness and vast vistas, the majesty of mountains, lakes and forests, the boundless herds of God's most magnificent creatures, created wonderment amongst the first explorers, settlers and administrators. This Eden, then, had to be protected, a perspective that empowered the colonizers to become its self-proclaimed defenders. Although what 'protection' meant was rather different from how we would understand it today, it fitted nicely with the class-based protection of hunting rights in Europe that went back to medieval times.[96] The colonialists were intent on separating lands of recreation – mainly big game hunting – from the uncivilized business of African life in general.

Access to Eden was controlled by the ideology of colonial hunting, and the lands in which wealthy and powerful white men could carry out that 'sport' became symbols of European dominance, a dominance defined by conceptions of class imported from Europe. An emphasis on their separation from and superiority over Africans helped the European settlers carve out the terrain of their conquests and fix themselves in their new homelands. For all the talk of the purity of relations with nature demanded by their hunting, the control of land and resources also secured very practical requirements, from revenues for the administrations of fledgling colonies, to ivory, hides and other commodities for trade, and even to meet the need for a supply of meat.

Although the notion of colonial hunting may have been an exercise in imperial power and control over the subjugated peoples of Africa and Asia, it was built firmly on the class structures that dominated Europe in the 19th century and which were in themselves tools of political and economic domination. The form of the hunting dictated by the ideals of the hunt was as much about society as it was about the activity itself, determining who could hunt, and how. It required strict acceptance of the responsibilities of the hunter to the hunted. The pursuit had to be fair, the hunter honest, and the kill clean. It demanded access to expensive hunting rifles and required an expensive licence.

Colonial hunting demanded the pure motivation of recreation rather than the base motivation of killing to provide sustenance – anathema to the colonial gentleman. For Africans, however, also engaged in their own rituals, it was efficiency that was important, and meat was the most direct, though certainly not the only, motivation. Naturally they, meeting neither the stringent personal demands of big game hunting, nor any belief in its values. were excluded from it – except of course as low-paid attendants: scouts, loaders, beaters and porters.

---

96 During the medieval period in England, being caught red-handed – with the blood of the king's deer on you – was a death sentence. In later centuries the wily poacher taking the lord's pheasants or rabbits was, if caught, equally at risk from the metal jaws of man-traps or deportation.

In Britain, class and power had over the centuries created 'wild' landscapes, such as grouse moors, stripped of working people. Now in Africa, big game hunting artificially created a mythic wilderness from which Africans were steadily excluded. They were barred from the hunt, and soon afterwards they were expelled from the land.

Nothing stays the same, though, and from the 1920s the Europeans in Africa began to regard wildlife differently. The Great White Hunters were transforming themselves into preservationists as the impact of their unconstrained killing had become hard to ignore; the carnage of the big game hunt no longer looked quite so pure. At least two ungulate species, the blaubok and the quagga, had already been driven to extinction at the southern tip of Africa, and the herds of other ungulates that, stretching in uncountable numbers to the horizon had entranced the early explorers, were gone. So then, rather than pursuing a vision of the perfect shoot or a landscape of leisure, Africa's white rulers drew on the American model of national parks to protect what was left of Eden.

Free of hunters – both African and white – and under the watchful eyes of reformed hunters, the great African bush would be protected. Class-based constructions of nature merged with the virtues of wilderness to set the direction for Africa's national parks. The first of these was, as mentioned earlier, created in 1925 in the then Belgian Congo, the second a year later in South Africa.[97] The legislation that preceded the creation of these parks and the others that would follow across the continent may have been designed to protect nature, but there were practical considerations too; the protection of large, highly visible mammals and the ranges they occupied legitimized the expropriation of vast areas of land.

The national park ideal and its application during the colonial period were firmly rooted in western constructions of nature. But conservation has much older traditions than these. The setting aside of lands, species and resources from the general course of human activity and the normal run of daily demands is not a new idea, nor one restricted to the west; people of all kinds and in all places have been doing it for as long as there are records to

---

97 The park in Congo was named after King Albert, and the park in South Africa after President Kruger. Thirty years later the British would follow the same tradition, naming Uganda's first national park after Queen Elizabeth II. Though President Kruger has been given credit for the conservation movement in South Africa, declaring the first nature reserve in 1898, it is not clear that either Albert or Elizabeth were advocates of nature conservation, and the names chosen seem more in line with the imperial practice of naming lakes, rivers, mountains and seemingly anything remarkable after the great and powerful.

record it, and no doubt much longer. And they have been doing it for as many and varied reasons as there are ways of understanding the world. People have protected animals, plants and places on the basis of beliefs that made it desirable or necessary to do so, using inducements, rules, regulations and sanctions.

Diverse ways of understanding the world and the human place in its workings have given rise to taboos, spiritual demands and practical prescriptions that one way or another protect nature. Claims on sentiment, morality or pragmatism have also prevented, restricted or controlled the use of nature. Hunting particular animals, cutting certain trees, grazing specified areas, fishing certain lakes or reefs or bays, have all been the subject of traditional regulations. Restrictions that may be applied to some individuals or groups but not to others may come into force during specified periods or seasons, or respond to the demands of different deities or spirits.

Many peoples set places aside from everyday use, reserving them for religious and spiritual functions. Throughout West Africa, priests perform rituals and prayers in sacred groves, and secret societies initiate the youth and teach them the mysteries of the forests and of life. Forests strung along the East African coast are homes of the spirits of tribal ancestors. Forests surrounding ancient Christian churches in Ethiopia protect trees that shelter the spirits of the dead. Pre-Christian tribes in Europe would worship in sacred groves and standing trees, and protect them. The royal forests of medieval England, the hunting parks of Chinese emperors and the organized hunts of Zulu chiefs and their warriors – all of these reserved the privilege of hunting to élites, thus conserving wildlife through the regulation of hunting, though this was not necessarily their purpose.[98]

Mburo provides an example of grazing resources set aside from everyday use.[99] The kings of the Banyankole controlled the area, famous for the quality of its grazing and its salt grasses. Mburo had access to permanent water as well, and, bounded to the east and south by lakes and wetlands, was easy to defend against raids from neighbouring tribes. These desirable characteristics meant that Mburo was kept as a special preserve for the royal herds. Though they were owned by the king, the herds of selected cattle of especial beauty were a kind of bovine national bank. They could be distributed in times of need, were a source of patronage, embodied the values of pastoral excellence, and were a source of national pride. Naturally these herds should have the best grazing, access to water and protection from raiding warriors. So of course other uses and other users had to be restricted.

The care that the Native Americans took of their lands and resources, in contrast to the behaviour of European settlers, have been well documented. Though founded on a fundamentally different worldview, the Native Americans' reverence for nature would seem to have offered up synergies with the ideals that stimulated the conservation

---

98 Hunting by English kings and their nobles signalled their power and authority as well as allowing them an activity they enjoyed. The Zulu chiefs did the same while cementing relations with and between their warriors.

99 There are many others around the world, particularly in arid areas where irregular rainfall requires grazing to be saved for emergencies. The Hema system of the Middle East is a good example of this.

movement. Similarly, as these conservation practices were extended around the world, there were opportunities to learn from and engage with ancient traditions and institutions to design initiatives with local resonance. Given the richness and diversity of traditional ways of engaging with the natural world and how effective they have been, it is surprising that they have made such a minimal impression on contemporary conservation practice.

The failure to integrate traditional forms of conservation, or apparently even to consider them, may be due to the assumption of western cultural superiority that was the norm at the time. Engaging with local cultures in a practical sense would also have run interference with the imposition of western values that were so helpful to the imperial enterprise. That little has changed in more recent times, even as conservation has been pursued by the institutions of independent nations, is more difficult to explain.

The early proponents of modern conservation in America were amazingly successful in promoting an idea that must have seemed contradictory to the primary values of the time. They were able, it seems, to ride the zeitgeist of the desire for a national identity that would be distinct from that of Europe. The demand to separate people and nature required the new parks to be explained in terms of nature's transcendent values, and the need to halt activities that diminished them. Hunting in the parks had to stop, as did the cutting of trees, the grazing of livestock, even the picking of flowers. Perhaps most significant for the future of the parks, it also meant that people could not live in them.[100]

It is not immediately obvious why the strict separation of the natural and human worlds was felt to be such an essential requirement for the parks. It may initially have been as much about the practicalities of defining and managing them as a reflection of the duality that lies at the core of western relations with nature. What is evident, though, is that this separation soon became an ideal, and the national park became the apogee of the western world's endeavours to protect nature.

Given the domination of western culture throughout the colonial era, and the marginalization and even conscious dismantling of relations with nature amongst traditional peoples, it is surprising that national parks have been taken up so enthusiastically by post-colonial governments and élites around the world. This is even more surprising when their exclusionary nature is so evident and so damaging to the immediate interests of so many. Yet during the latter part of the 20th century, the number of national parks gazetted increased dramatically.

---

100 Though these are the common rules, there are exceptions and anomalies. Hunting may be banned, but fishing is often allowed. Nobody may live in a park, except those charged with protecting it, but they often pick the very heart of the protected area to build their homes. These rules are presented as necessary to prevent the disturbance of nature – yet tourism, which often creates very high levels of disturbance with its campsites, hotels and busy roads, is promoted and encouraged.

Their adoption by the colonizing process may partly explain their current popularity, but parks would not have been so decidedly successful without some other, and universally appealing, characteristics. The idea of the national park has become virtually synonymous with the idea of protecting nature amongst the public and within the conservation community. This is so, despite the existence of many other ways to protect nature and even other kinds of legally protected areas. The concept of exclusion and separation continues to prove attractive and continues to be enthusiastically adopted by governments, notwithstanding the difficulties of establishing and managing them, as the history of Mburo demonstrates. Grafting other ideals onto the core concept of wilderness has helped make parks relevant, especially values attached to iconic species and the concept of biodiversity.

When I started work in conservation, it seemed inappropriate to talk about nature in emotional or personal terms. Yet the flood of new parks being declared around the world at about the same time were, in truth, responding to these kinds of values. Hidden beneath a cloud of scientific and economic myths, they were nonetheless based on emotional and culturally relative values. Representing parks in terms of science and economics, it was fondly believed, would provide more immediate, compelling and acceptable justifications than would extolling their beauty or spiritual worth.

Instead, we have built a house of cards by describing the economic contributions that parks make to communities, and we have papered over the cracks in the walls with other stories. We have fabricated a tale that to save ourselves we must save biological diversity, constructing an imperative that requires the protection of every species and each variety of every species. We have frightened the public and ourselves by implying that without our parks the very process of evolution itself will stop, that the ecosystems we depend on will fail, and that the world as we know it will end.

Scientists described these perils in scientific terms, economists described them in economic terms. All the plants and animals that made up nature were, in this story, a repository of vast wealth, and were also the providers of goods and services that could never be replaced. Lose them, and the wheels would fall off our national economies. Nature, writ large as biodiversity, became the myth to trump all others.

A practical opportunity must exist before a decision to establish a park is made, but illustrations of its purpose must lie at the heart of any pronouncement. Though rarely acknowledged, culture lies at the centre of the values picked to explain parks, just as it lies at the centre of all values. Instead, the multiple values of nature have been transformed into a single idea, the notion of biodiversity, and all the representations of nature's worth are now hung on this. This makes science the sole means by which nature can be understood and by which decisions about it can be made.

Sociologists, anthropologists and philosophers may describe a park as some kind of symbolic environment that gives meaning to nature. But this is not how conservationists see it. Our training in the natural sciences, operating through the hidden influences of western culture, conditions us to see parks and the natural world they protect as species, habitats and ecosystems, and all the complicated interactions between them and the physical environment. This is the world we are protecting, and we are protecting it from anything we identify as unnatural, and we define 'unnatural' as anything to do with human beings.

Separating the human from the natural allows us to regard humanity as distinct from all other life, and apparently subject to entirely different rules. It requires us to search for facts that demonstrate the uniqueness and separateness of humanity from the rest of nature. We propose and endlessly modify arguments to demonstrate this. 'Only humans are tool users' it was suggested, until chimpanzees were found using tools to get food and water or compete for dominance.[101] Well if chimpanzees use tools, certainly only humans *make* tools. But crows, not exactly our closest relatives, select, snip and modify twigs to probe for insects. This thinking allows, perhaps even requires, the separation of humans and nature to be reflected in idealized protected areas. The consequences that stem from imposing this dualism on parks is that all too often a gulf between the park and its neighbours opens up.

It is perhaps surprising that most of Africa's national parks have been declared since independence. During the 1970s the number tripled – and, just as in the colonial period, they were created by governments that simply declared them and cleared them. The use of power and force has a long history in conservation, even into current times, and though sovereign states take the lead, the role of western conservation groups and more recently, development agencies, cannot be ignored. Great pressure was, and continues to be, mounted on governments to protect wildlife and its habitats. Though the overt agenda of protecting wildlife may have been highlighted, Africa's post-colonial ruling classes seemed as concerned to establish their authority and power as the colonial authorities had ever been, and just as happy to expropriate land to achieve this.

Justifications for the protection of parks and wildlife made by African governments show a mixture of appeals to national heritage, international prestige and economics. Just prior to his nation's independence, the soon-to-be president of Tanzania, Julius Nyerere, gave a speech in which he argued strongly in support of conservation. 'The wild creatures, amid the wild places they inhabit,' he said, 'are not only important as sources of wonder and inspiration but are an integral part of our natural resources and of our future livelihood and well-being.' This statement (whether or not penned by a foreign conservationist, as

---

101 A particularly interesting use of a plant as a tool has been recorded in chimps. Chimps pluck leaves from certain bushes and after folding them carefully, swallow them whole. These leaves have a bristly surface on one side and, swallowed without chewing, help scour out intestinal parasites. Merely chewing up the leaves in the usual fashion would not do the job.

has been suggested), mixing romanticism, pragmatism and nationalism, is a good example of the kind of explanations that African leaders and their technocrats have advanced for adhering to a conservation ethic based so obviously on foreign ways of seeing nature. Protected areas reflect the constructions of nature of those who create them.

A park based on monistic thinking that recognises no distinction between the human and natural domains – if such a thing could be conceived of – would surely be very different from one based on a dualistic worldview. It would still, though, reflect the values of a particular group, a specific culture at a point in time, not something of universal value. The powerful place of thoughts, beliefs and ideas in the creation of a park is not generally considered; but protected areas are necessarily, like cultural landscapes, as much about their creators as about nature.

Modern conservation represents values that are often irrelevant to the ordinary person and may be contrary to their values, especially in developing countries where parks largely mirror the values of educated élites. Many developing countries have legal systems based on values that are not those of the majority, and that often conflict with laws based on traditional beliefs and cultural institutions. This is certainly the case for most conservation law. Community-centred conservation tries to overcome some of these conflicting beliefs and to change how national parks are perceived by local people and their managers.

National parks can, if wanted, embody a bundle of overlapping values and functions. Currently, though, they are constrained by the narrow understandings imposed on them by both laws and by conservationists. To understand national parks as they are today, we have to understand the values and intentions of their creators, managers and supporters. To understand parks as they could be requires us to recognize that nature is something different to different people, and that parks established to protect nature must therefore also represent something different to different people.

4

How to save a park

I drove up to Mburo on a bright rainy season afternoon in 1991, anvil clouds stacked high in a blue sky. I travelled through alternating bands of torrential rain that flooded water across the road in streams, and sun that dazzled and painted the landscape in primary colours. Thunder grumbled incessantly around the sky, lightning cutting around the storm clouds. I turned off the tar road, leaving the storms behind. Dripping trees and soaked pastures opened fresh and cool, washed clean, before me, smelling of organic life.

As the sun lowered towards the horizon colours deepened and saturated like an under-exposed slide, the puddled murram road, the wetted grey of tree trunks, the crimson glow of tiny hibiscus flowers on tall stems. The sun threw sparkles into the grass and hung them from leaves and on spider webs, and splashed gold into the pools of water standing in the meadows. It was three years since my last visit, that visit which, disquieting though it had been, had nonetheless inspired in me a spirit to resist the demise of the park from which had grown the idea for the project that had brought me back to Mburo again.

Alone this time, I had turned south off the main road at Akagati, a grimy settlement expanding rapidly on profits from the charcoal burning that was clearing the rangeland of its trees. The village took its name from the park's gate, now long gone, nearby. I bumped down a track through the government ranch of Nshara, slowing to navigate water-filled holes, and slithered over stretches of muddy corrugations that sent vibrations through my spine. Finding that I had entered the park at some point without knowing it I followed the vehicle tracks that wove around trees and termite hills, breaking into loose skeins to negotiate swampy patches and acacia thickets. I stopped to smell the pollen of the trees in full flower and listen to the rustling of damsel flies and butterflies massed along the edges of dams and in plashy corners. I watched white egrets stalking between the legs of cattle as they grazed.

The sun was approaching the line of hills across the lake as I climbed the gentle hill to the park headquarters. I passed through the same makeshift gate I had passed three years earlier, a rough branch resting on two cleft posts, the duty ranger in a ragged uniform half-waving, half-saluting, dragging it aside to allow me to enter. Some small brick structures

*Above: Grassy valleys in Mburo
frequently run with flood water,
attracting wildlife of all kinds*

*Left: Red hibiscus flowers show brightly
against the rainy season grasslands of Mburo*

had been built since my first visit, though most remained simple mud and pole structures. The semblance of a national park was vested largely in the ungainly radio masts and aerials, guyed with heavy wires that vibrated in the breeze. I drove past ranger lines, long low huts of mud thatched with grass and papyrus, passed a handful of incongruously decorated tourist bandas[102] scattered through a small olive and acacia grove,[103] and stopped at two metal uniports[104] joined by a covered walkway.

From this unprepossessing structure stepped Arthur Mugisha, the newly posted warden-in-charge, wearing a pressed khaki shirt, plastic sandals and a broad smile. We shook hands, my attempt to achieve the resounding clap and snap of a Ugandan handshake failing dismally. We had first met during the fishing village survey in Queen Elizabeth National Park. Arthur had been with me on that first visit to the park where we had together conceived the idea for a project to save Mburo. And now, against all odds, here I was to join Arthur to implement that project.

'Welcome to Rwonyo. Welcome back. Welcome home,' Arthur said. We had been planning this meeting for months, pushing so that we would both be stationed here together; and by strokes of both good and bad luck it had happened.

---

102 'Banda' means a small cabin or house, especially for tourists. Widely used in East Africa but not apparently known outside the region, bandas were constructed to give park workers or visitors more robust accommodation than that provided by tents.

103 I was surprised to find that one of Lake Mburo's characteristic trees is the olive, *Olea europaea*. Wild olive trees are found across Africa to the Cape, and across Asia all the way to China. It is the same species that has been cultivated for thousands of years throughout the Mediterranean region, but in sub-Saharan Africa only the wild variety is known and its small, shrivelled fruit are neither eaten nor used for oil.

104 Uniports are octagonal huts made of metal panels bolted together and placed on the bare earth or, if you are lucky, a wooden or concrete platform. Like ovens during the day and freezing at night, they seem to have been designed to make life uncomfortable.

This collection of rudimentary buildings, mainly constructed from materials taken from the park,[105] was basic but adequate for the task in hand. It met the modest needs of the staff posted to Mburo, and would have to meet mine too. Its significance, though, and the reason why the previous warden had worked so hard to establish it, was that the existence of the headquarters proved that the park was real, that Mburo and its wildlife had not been abandoned.[106] Rwonyo Park Headquarters would become my home, and for the next ten years I would work with Arthur, his staff and many others in the hope of ensuring Mburo's survival.

The young men who had joined me on my visit to Mburo three years before – had been as saddened as I to see the decline of the park, and we had agreed then and there that something needed to be done. Our determination to save Lake Mburo National Park was born, perhaps, of our ignorance and inexperience, but in my case from an immediate and strangely nostalgic longing to return there. I wrote the first notes of what would become a proposal for a project, in the uncertain light of a lantern, sitting by our campfire on the reed-fringed edge of the lake. I carried this briefest of outlines to Nairobi and the offices of the African Wildlife Foundation, an American conservation charity whose director I'd had the good fortune to meet. Our proposal seemed to fit with the foundation's thinking on the need for partnerships between parks and their neighbours.[107] A project designed to learn from what had gone so wrong at Mburo could help restore the park and set it

---

105 In the absence of cash, a great deal can be built with labour and natural resources. The previous warden had encouraged his rangers to build their own facilities. They made bricks from swamp mud mixed with papyrus stems and dried in the sun, and they laid them with more mud as mortar and covered them with a skin of cement to rainproof them. Roofs were constructed from branches, and thatched with papyrus from the wetlands or thatching grass, all harvested in the park. Unfortunately, the need for timber for windows and doors led to a local timber merchant being contracted to log the park's only patch of true forest, and most of the important trees were turned into timber. This small tragedy seemed to demonstrate an interesting perspective of conservation in Uganda at the time. Parks and game reserves had originally been established for large plains game animals and their predators. Forests were managed by the Forest Department for resources, not conservation. This allowed the warden, trained in the 1960s, to think of the park's only area of forest, without doubt the most biologically rich part of Mburo, as a source of revenue.

106 Uganda's tourism, closed down in the mid-1970s by Idi Amin, would take years to recover. The warden posted to Mburo on his return from self-imposed exile in Germany was Amoti Latif. Without funding from the cash-strapped National Parks he invited his German friends and connections in Kampala to visit him. First they came and camped, and then they stayed in the bandas he built for them with money they gave him. As well as generating a budget to rebuild the park, these first visitors demonstrated to the staff and neighbours that Mburo still survived and was not quite abandoned.

107 The African Wildlife Foundation is an American conservation charity. It was established in 1961 as the African Wildlife Leadership Foundation, with the aim of building a cadre of African conservation professionals. Its 'Parks – Neighbors as Partners' programme was a response to the growing acceptance of the need for the African public to support conservation. The simple idea I carried to them – which in the fullness of time became the Lake Mburo Community Conservation Project, and later, the Community Conservation for the Uganda Wildlife Authority – was a good fit for this endeavour.

Arthur, Francis and Tony, the Queen Elizabeth survey team who visited Lake Mburo in 1988 when we developed the idea of a project to save the park

on a sustainable course. It could also help the authorities learn how to work with local communities, rather than needlessly and endlessly battle against them.

The arcane language of project development with its logical frameworks, crafted statements of goals and objectives, designs for verifiable indicators of achievement, and details of multitudinous activities, targets and milestones was brought to bear on our simple and perhaps naïve idea to save Mburo. Our project concept was transformed into a fully fledged proposal, the Lake Mburo Community Conservation Project. It was two years, though, before the Swedish International Development Agency agreed to fund it. I was working for WWF International in its headquarters in Switzerland when I received the news. Would I, who had set this particular ball rolling, like to return to Mburo to implement the new project? It was the work of a millisecond for me decide yes.

If my time in Queen Elizabeth National Park, nearly a decade before, had been a childhood dream fulfilled, this new adventure was to be even better. I was no longer the young supplicant begging for a place at the table. I had some claims to experience and some thoughts of my own. Now, suddenly, I had a project to test them and myself against. I had a chance to play a larger part in saving the world, a chance to make my own impression, something I could put my energies into and achieve something I could put my name to.

Mburo was without doubt in need of saving, and I was going to save it. By good fortune it turned out to be me who was saved; saved from my unbounded but unfounded confidence in myself and my powers; saved from a failure to question or consider anything very much; and saved from a self-centred and self-serving agenda that would have sunk the project before it started. With gentle tact, the wardens who became my colleagues, partners and friends, Arthur, along with Moses Turyaho, the newly appointed community warden, put my feet on firmer ground. Schooled in the practical facts of life in Uganda, in the nicest possible way they set me straight. The reality of the situation we were intervening in needed layered and complementary tactics, all built on relationships, personal and institutional, local and national.

'Mark, we will help you, and you will help us. No gambling, no chancing. This is Uganda National Parks, and we will need Ugandan solutions to succeed.'

Arthur, Moses and I shared the work of implementing the project and managing the park. We swapped ideas, roles and responsibilities in a way that was not and could not have been designed for in the project. Without the happy accident of our being available, being so like-minded notwithstanding our different histories, experiences and knowledge, and being able and willing to devote our entire energies to a single mission, I don't believe we would have succeeded. We were not working alone, of course, and many others played essential roles, some working with us in Mburo, some in distant offices, and some implementing projects with complementary roles or delivering essential activities outside the brief of our project. But without the rapport that evolved between us three, I doubt the project would have persisted or the park been saved.

When I say the park was saved, knowing that this assertion can be challenged, I mean that the park became, again, and still is, a place in which nature is given precedence over the immediate material interests of the nation and the day-to-day activities of communities. When we started our work, however, this was not by any means the case, and the way to achieve it was far from clear.

Mburo had little support locally or nationally, its infrastructure was in tatters, and its staff demoralized. Meanwhile, its opponents worked secretly, and in some cases openly, against it. Assessing what looked like a pretty bleak situation, the three of us agreed we would work for five years, and if we felt this had got us the top of the hill, we would consolidate the achievement over another five years. At the first meeting with our donors, held on the crest of Rwonyo Hill in the old headquarters building we had restored and

*Villages and fields covered large parts of the
national park and threatened to undermine it*

equipped as an interpretation centre, the sound of long-horned cattle lowing from the valley below, we all accepted that ten years was a realistic time frame and committed ourselves to stay the journey.[108]

I am happy to believe that we reversed Mburo's steady decline and brought it back from the brink of dissolution. Our efforts had phases and stages; and we had reversals and failures. At the centre of the story lay relationships and partnerships. The achievement, if such it was, materialized over the years as we developed and implemented ideas that we felt were the right ones, and designed and undertook activities that would deliver on the many promises made to our partners, our donors and, especially, to our neighbours.

Our project was called the Lake Mburo Community Conservation Project, a name that seemed to signal its intent and set out the challenge, but which wrapped the endeavour in ambiguity. Were we conserving Mburo or the community? Was working with communities the object of the project, or a way of working that would save the park? What were we actually trying to achieve? Allowing this lack of clarity to persist was perhaps intentional, perhaps even necessary in the highly contested circumstances we

---

108 Three years was the standard project length at the time, so for a donor to commit to work for ten years was unusual. I think all parties realized that anything shorter was unlikely to succeed. As it turned out, the original donor, the Swedish International Development Agency, did not fulfil its promise, pulling out after the first grant had been spent. The African Wildlife Foundation was, however, as good as its word; and we were fortunate to get support for a second grant from the United States Agency for International Development.

were operating in. Sometimes different messages can be crafted for different parties to get them all going in the same direction, and the vagueness of what we were trying to achieve might have helped us with this.

One of our earliest tasks was to explain what 'community' meant in the context of the project and why community was at its core. Many of the park staff could only see Mburo and their efforts to protect it as entirely positive, all good. We needed them to see that Mburo was also problematic, especially for its neighbours. We needed the rangers and wardens to recognize that they had a responsibility to reduce negative impacts and look for ways for the park to produce positive outcomes for local communities instead. Equally, the staff needed to understand that the local communities could support the park, help it even if they were minded to, or undermine and oppose it if they were not. Their choice to support or oppose the park would depend on their experiences of the park and their perceptions of our efforts to protect it. The wider conservation community had embraced the realization that the interests of people living hard up against parks and reserves needed to be properly understood and considered. We had to build a similar understanding at Mburo and turn this theory into positive and practical actions.

Responding to the wave of enthusiasm for market-based approaches to solving the problems of the world and, encouraged by the sudden collapse of the communist Soviet Union, the conservation movement began to embrace a financialized vision of nature and its parks. Projects designed under the rubric of 'integrated conservation and development' embraced demands for 'wildlife to pay its way' by contributing to local development. Protected areas had a duty to support the livelihoods of their neighbouring communities and contribute towards the national economy.

Interventions were based on the assumption that communities were forced by poverty to destroy the natural resources on which they depended, making themselves poorer in the process. But when communities were made less poor, they would use lands and resources sustainably, breaking the pernicious cycle of degradation. Activities that integrated conservation and development were designed to reduce poverty and dependence on natural resources, creating space for people to embrace longer-term thinking and practices, thus relieving pressure on their local protected area.

Though there were, and still are, numerous difficulties with the overly simplistic assumptions that underpinned these approaches, and demonstrating successful interventions did not prove easy, the logic was so attractive, and the desire for what came to be labelled 'win–win' scenarios so strong, that these approaches came to dominate conservation programmes. The illusive nature of these much-pursued win–win outcomes suggests, however, that they were not quite real; when pursued, they tended to quietly vanish away, hardly the thing to base a global approach to conservation on.

Perhaps it was this that made us insist that our project should not be labelled an integrated conservation and development project. We had few expectations of easily discovering win–win outcomes. It was, in any case, not clear to us that the park's problems resulted from demands for access to its lands and resources from people facing poverty. Rather the problems seemed to stem from hostility towards the park as an institution and to conservation as an idea, at least as an idea as it had been revealed to those people.

This antipathy, which we saw both in the presence of the efforts made to undermine the park and in the absence of local political support, seemed rooted in the history of the conflict over Mburo. It was the negative relationships between the park and its neighbours that needed to be tackled. We had to engage with the facts of the park's creation, but we also had to look at how the park was being managed day to day, and especially at how the park staff engaged with their neighbours. Against some opposition we set aside the label of integrated conservation and development, directing the project instead to activities that would engage park and communities in positive ways. It would be essential, though, for park staff to commit to the project's community work, the thinking that underlay it, and the tools and practices that would deliver it.

When we took the project on it was, as projects must be in order to attract funding, a highly designed set of interlocking ideas and actions. Detailed objectives and interventions had been sewn together following discussions between the African Wildlife Foundation, the Uganda National Parks, and the donors. Even community representatives had been part of the process.

Activities fell roughly under three objectives: support to strengthen the park; assistance for neighbouring communities; and building constructive interactions between park and community. The three objectives had to be understood and delivered as an integrated whole. We rebuilt and re-equipped the park so that it could undertake its basic, traditional functions of protection, while in parallel we supported the community warden and his newly established team of community rangers.[109] Through carefully linked activities we also supported communities, helping them to engage with the park and benefit from micro-development projects.

---

109 One of the most difficult things for most projects is achieving sustainability. The idea that everything a project provides during its lifetime should, once the project has ended, be sustained by government, local communities or other partners seems correct, otherwise the efforts made and the money spent to establish something, whether a building, an institution, a capacity or a process, will be lost. But such sustainability is a hard thing to design for and even harder to achieve. For example, while it is fundamentally unsustainable for a project to employ its own staff to deliver activities and outputs, it is often hard, if not impossible, for government agencies to meet requirements for new staff to undertake new tasks.

For our project, then, it was agreed that Uganda National Parks would create the new positions within the Community Conservation Unit needed to implement the project and change how the park worked, but that the project would pay the personnel and provide the resources for them, then at the end of the project the new positions would become part of Mburo's establishment. I am glad to report that after our eight years on the project, this is precisely what happened.

A challenge that would dog the project despite our best efforts was that the different parties we worked with often perceived our actions as contradictory to their interests. For example, our support for communities was interpreted by some within the park as undermining conservation interests. At the same time, our support for the park was thought by many officials and community leaders to undermine their determination to deliver economic and social development.

Moses, Community Warden, with Mwesigye and Florence, founding members of Mburo's Community Conservation Team

Building the park

Arthur, Moses and I cut our teeth on the activities that we thought would be easy to implement and which would give visible and tangible results reasonably quickly. These turned out to be investments in park infrastructure, with tourism in mind. Government and most of those engaged in the development process regarded tourism as, if not the main purpose of parks, then the best and most practical way to demonstrate their worth. It was considered essential to demonstrate the value of Mburo as a tourist destination, which meant, before doing anything else, fixing its often-impassable roads. Without better roads there would be no tourism. But building roads was completely outside the scope and scale of our project. Fortunately, another project was building roads in the parks and was already busy in Queen Elizabeth and Murchison Falls. We lobbied hard, against some opposition, for the road crew to come to Mburo before they set off for Kidepo in the north, from where it was uncertain they would ever return.[110]

Our lobbying was successful. It was exciting when the great machines arrived to start the work of road-building, giant yellow graders with blades for carving the camber of murram roads, excavators and dumper trucks with wheels so massive that they dwarfed our vehicles, and fuel carriers and water bowsers, and all the paraphernalia of road building. It was shocking to see the disturbance that making even simple earth roads entailed, the

---

110 Kidepo Valley is located in a distant and inaccessible corner of Uganda. Once the machines were there we were unsure they would ever be returned to the west – not just in practical terms, but also because we even had some fears for the safety of the road crew. Kidepo lies in Karimoja. The Karimajong were, and still are, a fiercely independent tribe known for their cattle raiding. When in 1979, Karimajong warriors attacked the government armoury and armed themselves with automatic rifles, cattle raiding escalated into an unpredictable and dangerous mix of intertribal conflict and banditry carried out by heavily armed and unconstrained warriors.

digging and mounding, the bulldozing of piles of earth, the cutting of curving drainage channels. The roads, with their ditches and culverts were much wider than we had expected, but we quickly accepted the need for this. In addition, the park was surrounded by wetlands that were impassable during the rains. Raising and compacting banks of murram formed a drivable skin over the treacherous black swamp soil, but they needed constant maintenance. We had all spent too many hours sunk to our axles in stinking, swamp pits not to welcome the new roads, however much the disturbance to our quiet corner of the world.

Agreeing priorities and orientations for the roads with the engineers could be a bit of a challenge. The road team wanted to get in and out of Mburo fast – the reduction in the size of the park and the continued presence of farmers and cattle-keepers had led some to question whether Mburo was worth investing in at all – and they had a way of working that sometimes clashed with ours; whereas we were strongly focused on involving communities in everything we did, the engineers were focused on technical design and costs and speed, and they resisted our suggestion that they should discuss their plans with us and with the communities.

A carefully negotiated agreement had been reached with the fishing village at Rwonyo to relocate to a new site closer to the park boundary. But this was nearly derailed when the road crew opened a road right to the fish landing. With this in place the fishermen decided, not unreasonably, that they wanted to stay where they were. It took a great deal more careful negotiation by Arthur to keep the plan for the new village on track.

More serious was the failure to reach agreement with a landowner to open a murram pit on his land. The acrimonious argument that erupted between the farmer and the road

The fish landing, located in the centre of the park, was relocated to a site closer to the boundary and easier to access

crew threatened to undo the improved local relations that had arisen from the decision to build the main park access through community land rather than through the government ranch, as giving farming communities year-round access to the main road was an important improvement to their lives. But as a result of the argument the warden had to intervene to make peace again.

Such difficulties were dwarfed, though, by the positive impact the new roads had on both the park and the community. All of us could now get about, even during the rains when the old tracks dissolved into quagmires impassable even to four-wheel drive vehicles. Tourists could reach the park, even in saloon cars. New tracks opened up areas of the park that had been too remote or difficult to reach, making Mburo more attractive to investors looking for locations to build tourist lodges. Traffic through the villages demonstrated that Mburo had outside interest and support. The new road helped the communities, too, opening up their farms to traders and easing travel to the highway.

We were not shy to build on these developments, asserting that the park was for the first time on the cusp of delivering tangible benefits; once tourism picked up, the government's brave decision to support it would be vindicated. So, riding on the back of the road improvements, we invested our limited budget in tourist amenities.

We put up signposts, opened campsites, built latrines and dug rubbish pits protected with heavy lids to keep out the vervet monkeys and baboons. We improved the bandas, equipping them with beds and mosquito nets and hot showers. Our most ambitious project was a split-level dining shelter – we felt we couldn't in all honesty call it a restaurant – for which we commissioned a set of park-green plates, bowls and teacups with 'Rwonyo' hand-painted on each item. Staff were recruited, and trained to cook and serve simple meals.

Our efforts were not restricted to supporting tourism, though. We invested in making park management more effective. We bought vehicles, installed radios and communication

The dining shelter, an early investment of the project to improve tourist facilities in the park

masts,[111] and established ranger posts around the park to strengthen the contested but unavoidable aspects of traditional law enforcement. We bought uniforms for the rangers, boosting their morale, and equipped them with boots, tents and water bottles so that they could patrol effectively. We also invested heavily in education and environmental awareness, converting a lorry into a

---

111 Before mobile phone networks replaced them, radio communications were amongst the most essential but also one of the most problematic and expensive components of park infrastructure. It was their radios that, even though unreliable, enabled the wardens to talk to headquarters in Kampala, and the rangers and wardens to communicate with each other as they moved around the park.

bus for school and community visits, and working with teachers in local schools to establish wildlife clubs and develop teaching materials. Somewhat against my better judgement, thinking of it at the time as propaganda as much as education, we also constructed a 40-bed education centre for school visits, which continues to function today.[112]

Regaining authority to recover integrity

Perhaps the most difficult issue that had to be addressed was the loss of Mburo's integrity as a protected area. Putting aside the legal requirements of an area designated as a national park, the values for which Mburo had been established by general consensus – its large mammals and its wilderness and landscape values – were not being protected. This presented us with a dilemma; although the three of us had committed ourselves to saving the park, we saw how establishing it had damaged the interests of both communities and conservation.

We observed daily how failure to protect the park was eroding its natural values, and that outside the park these were rapidly disappearing, but I had deep reservations about the rightness, effectiveness and sustainability of parks that excluded their neighbours and their interests. Arthur's challenge, as warden in charge, was to recover a degree of authority and control over the park that would allow it to be protected, but without further alienating communities. We could envisage ways this could be achieved, and we prepared ourselves to undertake them, but we understood the inherent contradictions in our approach. The most difficult question to address was that of access; who could do what in the park, and for what reasons?

I believed that Mburo could avoid or at least reduce the problems caused by its exclusionary nature by agreeing to give our neighbours access to some of the park's resources. I felt that, though not formally allowed and even contrary to the ideal of the national park, certain forms and sustainable levels of use could be negotiated that would meet both park and community interests. There would surely be difficulties and disagreements, but the approach would provide a way to establish relationships between the park and its neighbours that were resonant with the history of the relations that had existed for centuries between the local people and the area.

The three of us could envisage the park's neighbours negotiating access to many of its resources including fish, meat, medicinal plants, papyrus, and even grazing and water,

---

112 Though I had reservations about our education programme generally, the proposed centre was strongly supported by everyone, especially the local communities. In fact, the centre has served the park and the community well for 20 years, despite a rather cavalier concern for its maintenance. The original centre was constructed with American government funds, and shortly before the time of writing was renovated and restored to its earlier glory, again with American support, and again under a project implemented by the African Wildlife Foundation.

but none of us could see a place for farming. This, we felt, had to be the bottom line for Mburo as a protected area. Farming might be part of British national parks, but in Britain there were no buffalos, baboons and bushpigs to contend with, any of which could wreck a farm overnight. As well as increasing conflict with wildlife, the farming system based on bananas and other crops required complete conversion of land, leaving little place for nature.

The challenge, then, was to regain control over the park without simply returning it to the standard exclusionary model. A delicate balancing act was required. Actions that built positive relations with our neighbours were balanced by simultaneously working to re-establish the integrity of the park. The notion of territory – problematic and old-fashioned perhaps – was inescapable at this point in the park's history and needed to be asserted.

During the late 1980s, following the collapse of the park, the wardens and rangers who had been given the unenviable posting to Mburo – it was a kind of punishment posting – had attempted to minimize their contact with their neighbours, especially those living within the park. This no doubt seemed the safest way to avoid open conflict. The pastoralists had been prevailed upon not to graze their herds along the main tracks that the occasional visitors used, and to leave a small area around the headquarters free of their cattle. This, it had been hoped, would create at least the impression for the tourists and officials that braved the broken roads to reach Rwonyo that they were in a conservation area. Otherwise, the park's pastoralists had been left to carry on largely as if they were outside the park. Similarly, the farming communities had been left to their own devices as they had planted their bananas, opened their fields, and built or rebuilt their houses.

The president's commitment to retain part of Mburo as a national park came with a caveat; those who had been living in the area before the expulsions of 1983 and 1984 had the right to return, and would not be forced out until land was found for them. That nobody was sure when this promise would be made good on created room for numerous areas of conflict. During the years after the fall of the park in 1985 and the start of our project in 1991, wardens posted to Mburo had dealt with this problem largely by avoiding it. Communities, especially the pastoralists, were not amenable to being told what they could do or where they could go as they waited for their land. Only the fishing communities had agreed to be gathered together in one place where they could be supported and supervised at the same time.

Arthur knew that unless some degree of control was regained over the park, and unless there was some acceptance of the authority of the warden, Mburo could not survive. In the early months of our project, Mburo seemed to shrink day by day, its wildlife diminishing and its beauty and tranquillity dissipating, while the villages and the cattle herds flourished. The interests of the Bahima and farming families living in the park lay in increasing the size of their herds, expanding the lands they farmed, increasing their population and building their communities. This strengthened their demands to remain – and, indeed, most were sure they would remain. In this situation, why would the Bahima bother to negotiate with Arthur or Moses on where they grazed their herds and dug their dams, or whether or not they would use poison to control ticks? They were able to do as they liked. And why would a farmer discuss anything with us at all, when all they needed to do was to sit tight as charities sank boreholes for them, churches were consecrated, and government constructed schools for their children?

It was clear that the longer things went on in this fashion, with uncontrolled use of land and resources, the more difficult it would be for us to establish any form of dialogue. Discussions could not start unless there was some acceptance in the community that the park was real, and that the warden had authority over its length and breadth. So at that point, what the objectives of having a dialogue might be, and what might be achieved for park and community from it were clearly less important to us than finding a way to initiate that dialogue.

The original design for the project, hashed out over the embers of our camp fire that night in 1988, emphasized the need to build positive relations between park and community, but a meeting point was needed before a dialogue could start. This was never going to be easy. Different principles would be needed to engage with the different groups if any degree of influence over their activities was to be achieved. Even to begin, to bring people to the table, the park needed local political support, and this was in short supply.

Arthur and I made strenuous efforts to present the park in a different light to local leaders, holding meetings with the member of parliament and the heads of local government departments. We even attended meetings of the Mbarara chapter of the Rotary Foundation.

In the end, our strongest card turned out to be President Museveni himself. His vision for the park might not have been clear, but his position was – he had ordered that one third of it was to be saved. His unannounced visit proved to be a tipping point for the park and the project. He arrived one afternoon and took Arthur into his vehicle, and together they toured the park. Arthur was able to explain his challenges and paint a picture of what he wanted to do. Though no formal communication followed this visit, news of it spread quickly. Arthur now felt able to push forward with steps to safeguard the park's immediate future. Though considered contentious by many, they were possible with the inferred backing of the president.

It strengthened Arthur and Moses' hand to act against the hunting and snaring that was rife inside and outside the park – agreeing to allow hunting of any kind was just not an option at the time. It allowed new arrivals settling in the park to be pushed out, and made it easier for the wardens to stop existing residents expanding their fields. Telling new arrivals and new fields from old was a challenge, but Moses' frequent visits to the villages helped. Arthur even felt strong enough to remove some of the herds that were permanently stationed in the park. Many of these cattle belonged to well-connected supporters of the president, soldiers who had fought with him during the long bush war from 1981 to 1986. It was a risky step to take, but the perception of presidential support allowed him to remove several large herds, returning a significant part of the park to the wild animals.

That the park had a project and even its own *muzungu*[113] also strengthened the hand of the staff, helping Mburo acquire the appearance of solidity and attainment, with new buildings going up and vehicles moving back and forth along the tracks and through the villages. Teams of social scientists from Makerere University surveyed neighbouring villages, asking questions, holding meetings and giving lunches. There were programmes to support schools and to arrange visits to the park; there were community development projects; and in time there would be initiatives to involve local leaders in the management of the park.

My presence behind the scenes was a clear indication that the money being spread around was coming from an internationally funded project and was not coming from the park's meagre resources. Moses could nonetheless argue that Mburo had brought something good for the communities, and suggest that if everyone found a way to play nicely together then more would come. The strategy of engaging communities to develop a more productive balance between exclusion and protection would not easily or quickly persuade those who had taken control of the park to give it up, but in time it brought them to the table. Building support amongst local leaders helped, but the promise of shared responsibility for the park was also attractive to our neighbours. It is a matter of real regret that this promise was never really delivered on.

Left: The Social Survey team gathered information in communities around Lake Mburo at the start of the project in 1991

Right: At work in my tented camp in Mburo, my base for nearly ten years

113 I was Mburo's resident white man, widely known and easily recognised. For me, there was no hiding.

Regaining and consolidating the integrity of the park required the project to commit to two major activities not envisaged in its design. The first entailed securing the park's territory. Park managers like to say that good fences make good neighbours,[114] ignoring the inherent contradiction of focusing attention on fences while expecting relations with neighbours to improve spontaneously.

But the case for establishing a new boundary for Mburo was compelling. Moses' struggle to retain the physical integrity of the park was made nigh-impossible when arguments over whether a newly cleared field or new dwelling was inside or outside the park were unresolvable because the boundaries of the park had not been agreed. Appeals to government eventually provided a formal statement on the size and approximate shape of the reduced national park. This meant surveyors could prepare a provisional map of the new boundary that Moses could use as the basis for negotiations with communities on the ground. Moses, his slight but wiry figure and unruly hair in endless energetic movement, spent months travelling the entire boundary, meeting with every affected community to agree the line the boundary would follow. Once each line had been agreed, the rangers, closely observed by villagers, would construct a concrete pillar and place it so that everyone – ranger, farmer and pastoralist – would know what had been agreed, and thus what was inside and what outside. It was an enormous job, but it made a profound difference, reducing our conflicts and arguments with villages overnight and demonstrating with assurance the footprint of the park. It might seem that establishing and marking the park boundary in this way indicated a retrogressive and backward-looking preoccupation with territory, designed to emphasize separation and bolster exclusion – but without it I doubt Mburo would still be a park today.

The second activity was to push for the president's promise of land to be met. An institution called the Ranch Restructuring Board had been established to identify land that could be allocated to landless pastoralists. The private ranches, from the 1960s scheme, bordering the park were assessed. If they were not being actively invested in and managed productively, the Board demanded that part of them be given up. This land was then allocated to pastoralists living in the park in relation to the size of their herds. Arthur stepped in to help with the head count – an exceedingly tricky task, in that the pastoralists were torn between hiding their animals for fear of being taxed and borrowing animals in order to secure a larger allocation of land.

---

114 This phrase, commonly used by protected area managers to justify their focus on boundaries and demands for exclusion, comes to us from the poem 'Mending Wall' by Robert Frost. Published in 1914, it would have been available to the first wave of park managers. It is noteworthy, though, that the words, 'Good fences make good neighbours' are not those of the poet but of his neighbour, who states them twice. The voice of the poet, also spoken twice, considers that 'Something there is that doesn't love a wall.'

Left: Moses Turyaho surveys the park boundary in preparation for building permanent markers to delineate the boundaries of the park

Right: Government officers count the assembled herds inside Lake Mburo to allocate land outside the park to the herders

A group portrait, to commemorate the great cattle muster

Arthur also set out to lobby the Board to include the farmers of the park under its mandate as well. Government officials came and valued their buildings, crops and trees, and the project set aside funds to compensate the farmers for these assets. This would provide them with money to invest in the lands they were to be given, a condition of the Board. On agreeing to leave the park, the farmers would be helped to open bank accounts in the local town, and when they did eventually leave, they were indeed paid. All of this took some time to achieve, but since the day in 1994 when the last farmer moved to his new land, Mburo has had no residents, legal or illegal, except of course for the park staff and the occasional itinerant project manager.

Talking to the neighbours

Moses, meanwhile, had led the initiative to engage our neighbours in dialogue rather than conflict. It was not easy. Households inside the park whose demands to remain on their land had not been accepted were, unsurprisingly, almost impossible to engage. Families outside the park had undisputed ownership of their lands, but even here, dislike of the park and its officers was strong. Moses represented the first real effort to reach out and interact with people in ways that were not simply about enforcing the law. He was the fresh face of a more open, consultative and participatory approach to park management. Behind the wheel of his Landcruiser he crisscrossed the village lands bordering the park in pursuit of conservation.

As he passed through dispersed settlements and trading centres scattered along the narrow tracks that wound through the open bush and closed banana plantations, he would see the road empty in front of him. Men and women, old and young, would vanish into houses, shops and fields, reappearing as he passed, occasionally urging him on with thrown insults and sometimes stones. Rumours reached him of plans to ambush him at some lonely corner, but he persisted. He asked local leaders to meet with him; he invited himself into the houses of teachers and parish chiefs, gently insisting that people hear him out.

Little by little he was given time to sit and drink milky chai or banana beer and talk, and his high-pitched, persistently enthusiastic voice could be heard through the windows of huts, brick houses and school blocks and from wooden benches set under shady mango trees. Frowns and stony stares gave way to waves as he passed by, and soon he was fielding requests for help.

Familiarity did not breed contempt but, instead, the beginnings of engagement and the glimmerings of the potential for cooperation. Though far from addressing the deeper issues of governance, especially the question of who made the decisions about how the park was managed, Moses' work in breaking down the anger, resentment and suspicion was essential and the beginning of what would follow. In time it would lead to the back and forth of discussion, the early stages of dialogue, and the desire to consult and be consulted. None of this was in any real sense the sharing of authority, or a true collaboration, but it was a start.

The park brings you benefits

Now that Moses was able to sit down and talk with communities, the design and implementation of our micro-development projects could begin. We called these projects SCIPs, which conjured up for me the slightly silly but positive image of our skipping together, hand in hand, through all our differences and difficulties – it actually stood for Support for Community-Initiated Projects. Though initially we had expected that

SCIPs would demonstrate the much-touted benefits of the park and deliver those much sought-after win-wins, this proved difficult. When it came down to it, we couldn't identify significant ways for our neighbours to gain direct economic benefits from Mburo other than through the hunting and grazing that so many of them were already secretly engaged in. Despite our efforts to identify and support projects that would show the park providing tangible benefits – ideally income or jobs linked directly to conservation – this disarmingly simple idea proved surprisingly difficult to achieve.

We soon had to accept that most SCIPs would fall into one of two camps: either projects to build social infrastructure, or projects to fund small enterprises such as bee-keeping or bread-making. These we could describe as 'conservation friendly' as long as they were not obviously the opposite. Our original project had been designed to ensure that the micro-developments we supported would demonstrate that the benefits that the communities received were linked to the conservation of the park, but it turned out differently; although we could easily demonstrate that the SCIPs benefited the communities, we could not show that the SCIPs benefited the park.

This was in part because I was keen to draw on best practice in community development and demonstrate our participatory credentials. The project fielded a team of social scientists who undertook detailed assessments of community needs and asked villagers to identify problems the park could help them with.[115] Such assessments were considered key to integrated conservation and development approaches, and though they might have been necessary and even unavoidable, they were also problematic. The needs of the farming and pastoral communities around Mburo were great, far beyond the capacity of the park to respond to, even with the funds available through our project.

To give an idea of the scale of the park as an operation at that time, its budget was so limited that for a couple of years the project doubled it by funding the making and selling of T-shirts. The danger of asking people what their needs are yet being unable to meet them seems obvious, but it was difficult to escape when the conservation model was so focused on demonstrating the economic value of conservation.

The next difficulty was that, when asked, communities did not request help to start tourism businesses, or help to improve their farms in order to supply the new tented camp being established in the park, or even to train to be rangers or guides or cooks. Instead, they asked us to build clinics and schools and better roads. Moses and his community rangers were soon busy in the parishes around the park, working with local leaders and

---

115 The project had been designed to restructure relationships between the park and its neighbours, not establish relations with the community itself. But the community knew about the project – the *muzungu* they saw driving in and out of the park was a dead giveaway – and they wanted to engage directly with me to agree how the project would help them. I had to bat requests for help over to Moses, and I tried to give the project as low a profile as possible. I annoyed my bosses and donors by failing to put up the expected logos and stickers on our vehicles or on the walls of the schools or clinics we helped to build. The community had to accept that the benefits they received came from the park, or at least that they came as a result of the values of the park; and they needed to develop relations with the park and its wardens, not with the project and its *muzungu*.

community groups to identify and design such social infrastructure projects, even though it was not at all clear that these SCIPs really helped the park.

The understanding behind the integrated conservation and development programme was that poverty led communities to threaten parks, either by opposing them or by damaging them, and that providing alternative livelihoods would make hunting or charcoal-burning unnecessary, reducing conflict. Investments in schools and clinics would have only long-term impacts on

Working with communities; a project that supported honey production using local hives

poverty at best, but the needs assessments we undertook showed overwhelming demand for such projects. Our desire to consult with our neighbours meant we had little choice over what we funded. We had done what was expected of us, we had laid out the terms of engagement, and we had to abide by the results. Our thinking about these micro-development projects had to change.

Rather than espousing their direct impact on poverty, we argued for the importance of these projects in terms of park–community relations. They were demonstrations of the park acting and being seen to act as a good neighbour. Good neighbours help each other where they can. We would do what we could to improve schools and clinics, hoping in turn that our neighbours would help us with our problems. Dealing with the continuing efforts being made to undermine the park throughout the first years of the project, the expansion of settlements, and, of course, poaching and grazing, all needed neighbourly assistance. Wishing to be good neighbours did not mean that either side stopped trying to manipulate the other – there was not enough trust or transparency for that. The communities were focused, perhaps with justification, on deals that seemed the most advantageous to them. Equally, however, it could be argued that the entire SCIP programme was designed to coerce poor and powerless communities to support the park.

After some rather difficult conversations with our managers and donors, it was agreed the project could support social infrastructure so long as it was made clear that the SCIPs were only being made available to communities because government wanted to protect Mburo and its wildlife.[116]

---

116 At a larger scale and with more resting on it, a similar discussion was going on with the Bwindi Mgahinga Trust Fund, set up and endowed by the World Bank to conserve Uganda's mountain gorilla. The trust, which funded community projects, had to revise its rules to allow it to invest in the social infrastructure the communities were asking for rather than the enterprises the trust had been developed to fund. Since then the pendulum has swung back and forth between support for social infrastructure and investing in 'conservation friendly' enterprises. Efforts to assess which of these approaches are most effective in building positive relations have yet to give a clear answer. Both seem to have their values and their limitations.

SCIPs might have taken many forms, but they all bettered the lot of communities and built positive relations between the park and its neighbours. They did not, however, change the fundamentals of power relations between government-supported conservation and community. Nor did they deliver to communities any real economic benefits generated by conserving nature.

Moses was determined that communities should contribute to their projects. He argued that if we paid for everything there would be less support to go around – the budget for these projects were not great. He also believed that communities needed to contribute if the SCIPs were to become theirs, rather than remain essentially projects of the park. If the park built a school and the community had no hand in it, would they really take responsibility for it in the future?[117] Moses worked hard to negotiate deals in which the park would buy the cement, timber and roofing sheets that were beyond the capacity of communities to pay for, and cover the fees of, the contractor, while the community would provide the bricks, which they could make themselves, and the labour.

They were sensible arrangements and founded on an important principle, but they did not always work as expected, as I found when I visited a classroom block we had helped build. A lesson was going on and I was led up the fresh brick steps to the classroom door. Stepping through it I found myself teetering on an edge a metre above the rough, steeply sloping earth floor of the classroom, looking down on the heads of the nearest students at their wooden desks below. There were no steps on the inside, so I made a rather ungainly entrance, scrambling down the rickety wooden ladder provided.

The floor of the classroom seemed to have sunk below ground level, and it took me a while to comprehend what had happened. The community had agreed to provide the labour for the construction but rather than digging into the sloping site to level the foundations, they had persuaded the contractor to build the walls upwards, perhaps with the idea that once the walls had been finished they would backfill the large wedge-shaped hole inside them. Somehow that didn't happen, and students, teachers and unwary visitors had to climb down and up the wobbly ladder, and endure a floor that sloped like a trick house in a funfair.

---

117 We experienced an example of what happened when there was no community input or ownership of a project. During a parliamentary election the local member of parliament came to us requesting that we roof the school at the local trading centre. We declined, explaining that the community had refused to contribute to the work that needed to be done. In some indignation the MP bought the materials himself and organized for the new roof to be constructed. A week or so later we saw the new iron sheeting gleaming on a fine new roof. A week later it was gone, the whole structure whipped off and destroyed during a storm. The MP had hired a contractor but was too busy campaigning to supervise the work done. And nobody in the community had either. The roof was so badly made that the first storm tore it off. A few months later we made an agreement with the community to roof the school with their labour and supervision. That roof remained in place. I am not sure if that community voted for the MP or not, but he was elected and remained a good friend of the park.

Though most SCIPs ended up building social infrastructure, we still sought projects that would establish economic links between the community and the park. Not only were they central to the market-based rhetoric we were operating under, but also they seemed to make sense in the context of the poverty of the communities in which we were working. They also seemed to promise enduring bridges between park and community. In the language of the time, we were seeking relationships based on 'enlightened self-interest': the park would deliver benefits to communities; communities would see they were generated by the park; communities would see it was in their best interests to protect the park.

We tried various approaches and invested in numerous activities, but in the end few could in truth be described as successful. Several SCIPs were based around tourism, as many efforts to demonstrate the local benefits of conservation areas are. We trained local people to act as tourist guides, but there were too few tourists to make it a viable livelihood option for them. Worse, we found the park rangers considered these new community guides as competition for the scarce opportunities to earn a bit of extra cash.

We designed tourist experiences that community members could deliver. One was a boat trip to search the swamps for sitatunga, the splayed-hoof relative of bushbucks that abound but are rarely seen, and the elusive papyrus gonolek with its pillar-box red chest and golden cap. It was an exciting venture – but demand was low and enthusiasm for the venture dissipated. The guides could not sit all day waiting for tourists that didn't arrive. They needed to be busy with other things. So when the occasional tourist did arrive, on being directed to the boat landing from the park gate they would find no boat and no guide.[118]

We tried to help local farmers to supply our dining shelter and the new tourist camps with fruit, vegetables, eggs and meat. But the ventures were too small and demand too irregular for local producers to make a go of it. It was also difficult for them to meet the requirements for the kinds and quality of foodstuffs expected by international tourists. For the tour operators it was more convenient to bring supplies from Kampala. Similarly, the thought that tourist camps would offer jobs to local people if we provided training was not realized. The jobs were few anyway, and the skills needed were too specialized for the sons and daughters of the local farmers and pastoralists. This would change in the future, but back then it was far easier for the tour operators to recruit camp staff in Kampala where skilled English-speaking cooks, waiters, drivers and guides could be found.[119]

---

118 In these days of mobile phones and the ease of communicating at the push of a button, it is hard to imagine how difficult it was to organize simple things just a few years ago. This would not happen today as those organizing the trip and those guiding it would be able to talk freely to make the arrangements.

119 The lack of skills is a common barrier to communities gaining significant benefits from tourism and takes long-term investment and commitment to overcome. Mihingo Lodge, established on the eastern boundary of Mburo, has worked hard to train local people. But though it employs all its staffs from the local community, none are drawn from the Bahima families that occupy the land immediately around the lodge. Staffs come from more distant trading centres where education levels are higher and more English is spoken.

When we realised that the park and the fledgling tourist industry could not support businesses or create local jobs, we redirected our attention to businesses to serve local needs. Such businesses might not build a direct link between the local economy and the park, but they could improve livelihoods – and that, according to the integrated conservation and development concept, would help. We worked with farmers to plant hedges of the fabulously prickly Mauritius thorn around fields to reduce damage by wild animals[120] We helped improve farming methods to increase yields, introduced agroforestry techniques, and investigated new crops and markets. We trained community groups in bee-keeping and honey production and set up small bakeries. We looked at the fishing industry on Lake Mburo and how it could benefit nearby villages.[121] In short, we were very busy and very active and, at the time, altogether rather sure our efforts would yield results.

Though our programme achieved some success, at least in terms of the interactions between the park and its neighbours, if less so in delivering any great benefits to communities, we also created a problem. High levels of participation are the hallmark of community conservation approaches. To succeed we had to engage closely with communities, gather information and perspectives from the members, and involve them in planning and implementing activities. But farmers and their families are busy folk, and expecting them to attend meetings, engage in assessments and join our committees without providing at least a meal and a drink seemed unreasonable. Meals and drinks soon expanded to compensation for the day spent away from their fields, which also seemed reasonable. At that point the project was funding the participation of our neighbours, even in the processes that would be directly beneficial to them.

Local officials are busy people too, as are government officers, and they too requested travel allowances, accommodation and finally, incentives to support the work of the project. Our intentions might not have been wrong, but the unintended consequences have been pernicious, creating expectations that are detrimental to both community and official action. As we found out, paying government officials and communities to work with the project meant that no one turned up to meetings unless paid to do so, even when it was part of their job or when they would benefit from attending.

---

120 The use of Mauritius thorn, *Caesalpinia decapetala*, was not without controversy, as it is an exotic species that needs close management to avoid it becoming invasive. When planted as a hedge and carefully trimmed, it grows into a barrier so thick and so thorny that according to farmers, not even snakes can pass through. Preventing bushpigs and baboons getting into their fields made a big difference to their crop yields. But maintenance was hard work, and the hedges took up too much of their fields to become very popular.

121 Management of the fishing industry was not included under the process of negotiating access to park resources because although Lake Mburo lay entirely within the park, it was the responsibility of the Fisheries Department to manage the fishing. The park could, however, and did, play a role in establishing spawning and nursery zones, relocating the fish landing site from the east to the west of the lake to shorten the distance to the park boundary, and looking at ways to manage access to fishing licenses to improve conservation outcomes.

Sharing the park

As the SCIPs had failed to build significant levels of economic connection between the park and its neighbours, we were still left in pursuit of those elusive win–wins. We began to explore options for negotiating access to the park's natural resources, which, we reasoned, would make more direct contributions to the local economy. We expected this to be difficult and contentious – after all, one of the fundamental tenants of the national park ideal was the strict protection of animals and plants. It turned out to be surprisingly easy.

It was quickly agreed that the law could be interpreted to allow local use of resources if this helped to strengthen the park and achieve conservation. Whether this demonstrated that the directors of Uganda National Parks at the time were visionaries, well ahead of their time, or showed the dependence of a weakened institution on foreign donors who had begun to demand such community-based approaches, is unanswerable. It is not likely, though, that similar agreements would be entered into today, and the resource use agreements negotiated in some parks at that time have been allowed to erode since.

Though Mburo had since the 1930s been under one or another form of government control, when our work began in 1991 it was still hard for communities to accept they were prevented from using its resources. That animals such as bushpigs and buffalos emerged from the park at night to destroy crops and threaten lives made it even harder to accept. And that they were not allowed to pursue the destroying animals when they fled back to the park was like rubbing salt into the wound. There might be different ways to look at the losses Mburo levied on its neighbours – and many conservationists denied any responsibility – but there was no arguing with the fact that communities suffered costs. It was equally impossible not to see that they blamed the parks. When Arthur and I had first conceived the idea of a project to save Mburo, we recognized the losses that national parks could levy on local communities, and rebalancing that equation and even delivering gains was central to our thinking.

Wardens of national parks in Uganda had some local flexibility,[122] which Arthur energetically embraced. But Mburo still had to operate under laws that demanded the exclusion of people and unambiguously outlawed farming, grazing and hunting. These three were therefore off the table when the talking about resource access began. Different groups responded to this status quo in different ways. The farming communities accepted

---

122 This flexibility was formalized in the Uganda Wildlife Act of 1996 which integrated Uganda National Parks and the Game Department under the Uganda Wildlife Authority. Despite many important changes and innovations, the exclusionary ideal of the national park was retained.

that they would not be allowed to farm in the park. They supported the concept of private land, which strengthened their efforts to secure ownership of their own, so they could accept government's claims to own the park. Hunting had been banned since colonial times, so nobody was going to demand that – and anyhow, those determined to hunt were just going to carry on hunting.[123] The issue of grazing was thornier.

The steady increase in regulation in Mburo – and, finally, the banning of grazing there when the park was created – was abhorrent to the pastoralists. Equally, the idea of domestic animals in the park was anathema to the authorities who took their lead from the wider conservation movement. All sorts of reasons to justify the exclusion of livestock were advanced: diseases would spread from livestock to wild animals; domestic animals did not evolve in Africa and unlike the otherwise rather similar antelope and buffalo, destroyed vegetation and eroded soils; tourists would shun parks full of cattle. Their resistance, though, was as much an emotional response to idealized views of national parks as it was about ecological or financial realities.

Taking our lead from Bwindi National Park, which in 1992 started sharing some natural resources with communities, in 1993 Arthur started negotiating local access to a few resources, starting with catfish.[124] During the rains, the open valley bottoms that joined the lakes and swamps flood. Catfish swim upstream into the flow of nutrients that are washing down the inundated valleys, searching for places to spawn. As the floods subside, thousands of them become trapped in shrinking pools. Herons, storks, kites and buzzards, leopards and hyenas too, descend on the threshing mass for an easy meal. People would also join the rush. Carrying open-ended conical baskets, they would thrust these down into the roiling water and reach down through the open top to grab fish after fish. Agreement to give access to this seasonal bonanza was easily reached and was an early achievement of the community conservation approach.

Other agreements followed where mutual interests and benefits could be accommo-dated. Parts of the park's extensive papyrus swamps were opened for periodic harvest, more than meeting the communities' requirements for mats, thatching and fishing net floats. An assessment of medicinal plants found that those in the highest demand were widespread, often growing as weeds in cultivated fields. This eased agreement for access to the few rare plants restricted to the park. Herbalists from neighbouring villages were

---

123 Though there was licensed hunting in Uganda until the 1970s, Idi Amin banned all hunting in 1978, and his decree has never been repealed. This means that hunting is not just illegal in protected areas like Mburo but is illegal throughout the country. The only exceptions to this were a small number of experimental safari hunting ventures, including one established in the ranches around Mburo.

124 Lake Mburo had had a thriving fishing industry, probably since the times of the Bachwezi Empire and certainly since the British had encouraged commercial fishing of the lake. But as the Ugandan lakes and their fish landing sites fell under the Department of Fisheries, even when they lay within a national park they were not considered as park resources. This was ironic, because the value of the Mburo fishery dwarfed the value of tourism or other park resources at the time, and because the management of the fishery had significant impacts on the park. The virtual disappearance of pelicans and the decline in fish eagles, for example, has been attributed to the level of fishing, while crocodiles are frequently tangled in nets and killed.

registered, and harvested controlled quantities of medicinal bark and roots as well as collecting seeds and seedlings for nurseries and plantations outside the park.

The communities approved of these agreements, as they responded directly to the people's needs, and softened the hardest edges of strict park management. Perhaps most importantly, the agreements demonstrated that communities' needs were being taken seriously and their members' interests were being considered. The agreements were also acceptable to park staff, as they indicated that although some compromises had been made, the wardens were still managing Mburo for conservation, and they were still in control.

In terms of the integrated conservation and development thinking that was beginning to dominate conservation practice in the 1990s, these agreements showed the park contributing directly to local livelihoods. Unfortunately, this was only part of the financial equation. Although it was hard to argue that Mburo was an engine of local development, its contributions to livelihoods were real, even though they were not sufficient to cover the equally real costs that Mburo levied on its neighbours. Mburo could meet neither the costs of the lost opportunities for the local people, nor even the direct costs to them of the crop damage they suffered and the loss of their livestock to leopards and hyenas – costs that were higher than the financial value of these resources.

Nonetheless, I was excited by the resource use agreements, as they seemed to demonstrate a fundamentally different way of managing relations between the parks and their neighbours. These agreements would establish and maintain connections based on shared interests and they would also, I believed, build shared understandings of what Mburo could be, and what it could provide to humanity in general and local people in particular.

A concern that integrated conservation and development sought to respond to was the economic dependence of communities on resources in need of protection. Such dependences were understood as a problem, a source of conflict that needed to be removed. Fast-growing trees were planted outside parks so that communities would not be forced to search for fuel or building wood inside them. Our support for growing medicinal plants outside Mburo was an attempt to meet community needs while reducing their dependence on the park.

There are certainly situations where parks cannot meet needs. The demand may be too high to be met by the limited resource; the resource may be considered too precious to be used at all; or the process of harvesting may be too disruptive to other values. In these circumstances, I could see that these 'replacement' or 'de-linking' activities were useful in reducing pressure on parks. In other circumstances, though, I saw a danger in de-linking people and their livelihoods from their local park, and felt that retaining links between them was important, not necessarily a dependence but an ongoing connection.

The resource access agreements at Mburo helped maintain direct, material and personal connections with the park and its resources, and retained the culture and practices of harvesting and using resources that kept connections to nature alive. Seeking to construct positive relations between parks and people and remove negative ones was important, but if the result was no connections at all, if all needs were to be met without reference to parks, why should people care about them?

Most of the resource use agreements at Mburo were made with the farming communities. Access to grazing and water was all the pastoralists asked for; they had no interest in the other resources on offer. The Bahima argued for it, too, on grounds of poverty and the wellbeing of their herds. They insisted that without this access neither they nor their cattle could survive. That they were being squeezed out of the ranches bordering the park made their situation even worse. They reminded us that Mburo was the heartland of the ancient Nkoro kingdom, that they tended cattle on behalf of the nation and the ancestors as well as themselves, and that government was obviously mistaken in excluding them in favour of hyenas and jackals.

The Bahima were up against one of the strongest and most deeply embedded prohibitions of parks, the prohibition against domestic animals – especially, in practice, the animals of pastoralists. In Africa the grazing of cattle inside parks was abhorrent to modern conservation and its institutions. Furthermore, governments of all stripes had long viewed nomadic pastoralists, with their apparently endless and aimless perambulations and their supposed obsession with the number rather than quality of their animals, as worthless, irrational and uncooperative.[125]

But even putting these perspectives aside, it is easy to see why it was hard to discuss grazing in Mburo. Not only was it counter to long tradition of national parks, undermining accepted conservation wisdom and against the law, but more immediately to the point there were upward of 20,000 head of cattle there already. Even for a farm trying to do no more than maximize production, this was far above sustainable stocking rates. The pastures were so heavily grazed that even during the rains herds had to be driven out of the park to find grass. If Mburo was to meet any of its conservation aims, there needed to be far fewer of these bovines. Finding a way to remove them from the park was not going to be made any easier by agreeing to take in more.

After long and hard negotiations, it was accepted that if dams and reservoirs outside the park dried up, the herds could be driven down approved corridors to the river and lakes, and then return to their grazing areas outside the park. The corridors would need

---

125 A great deal of research has demonstrated that in fact nomadic or transhumant pastoralists make highly informed decisions on the management of their herds and the rangelands they use, allowing them to make the most efficient use of marginal lands where rainfall is patchy and uncertain. Movement, sometimes over long distances, is required to achieve this. Unfortunately, although the application of western principles of range management to such lands and resources has been thoroughly debunked, governments, departments of agriculture and international donors continue to castigate pastoral nomads, describe their practices as damaging, irrational and backward, and put enormous pressure on them to reduce their herds and settle down.

to be wide enough for the cattle to feed as they moved to and fro, as otherwise the trek would be too demanding, even for long-legged, drought-tolerant Ankole cattle. Of all the agreements between the park and communities, only this one remained informal, a kind of gentleman's agreement between parties that continued to disagree in principle but found a way to move forward on some fronts nonetheless.[126]

Working with our neighbours

Governance is a relatively new word in the lexicon of conservation, though it lies at the centre of protected area practice. Governance is concerned with how decisions are made and by whom. It is about power, who holds it, and whether and how it is shared. Governance is in essence about who calls the shots. These questions were little considered when I first started working in Uganda in 1981, and even after the disastrous way the park had been created, there was little engagement with these fundamental concerns.

The Lake Mburo Community Conservation Project hoped to change the way the authorities interacted with their neighbours and how communities thought about Mburo and its managers. But we didn't attempt to address how the decisions regarding the management of the park were reached.

In retrospect, this might seem surprising; clearly, the way the decisions are reached, and the roles the parties play in making these decisions, are bound to affect how the parks are looked at by those whose lives are affected by them. But our project was a product of its time, and had to operate within its time. Influencing how Mburo was perceived was a goal of the project, but in 1991, if we had suggested the authorities share power with farmers and pastoralists, the idea would have been rejected out of hand – and, no doubt, the project shortly after, because the idea of asking communities how parks should be managed would be like putting customers in charge of banks, or foxes in charge of hen houses. Only educated professionals had the knowledge and perspectives needed to protect nature. Communities had no interest in wildlife beyond eating it. Given an inch they would take a mile, and would soon be demanding to hunt, graze and farm the parks. Given authority, communities and their advocates would pack Mburo's wildlife off, take it somewhere else, sell it to zoos and take the land.[127]

---

126 Lake Mburo National Park's first management plan, published in 1994, set the standard for consulting communities, with nearly 450 community members attending meetings to discuss it. The planning team also met community leaders, local and district government officials, central government officials, scientists, tourists and tour operators. The plan that resulted reflected and formalized the community approach, detailing management actions to establish sustainable resource use areas, support the Park Management Advisory Committee, implement revenue sharing, involve communities in tourism, and undertake community outreach programmes. However, the proposal to formalize the cattle corridors was rejected by the high-level consultation called to ratify the draft plan. Technocrats and government officials dominated that meeting, and the voice of the community was hardly heard. As a result the plan calls for the strict exclusion of cattle and for the provision of water sources outside the park; and there is no mention of the agreed corridors.

127 I have certainly heard all of these suggestions being made by community members, so the fears of park authorities are not without some foundation. Unfortunately, their response is not the right one.

It was also perceived that only the firm hand of government's loyal servants prevented Mburo from being overrun completely – ignoring for the moment that in 1985 it had in fact been overrun – and that their hand could only remain firm by being clearly and uncompromisingly in control. Though not all these prejudices about community attitudes to wildlife and parks were entirely wrong, the idea of sharing power was rejected because those who held it feared losing it. Looking towards Mburo at that moment, the authorities could point to an example of what happened to conservation interests when governments lost control.

The reason I was disposed to see things differently might have been that in England conservation could only be achieved if the interests of government, landowners and the public were reconciled in the design and management of the parks. So although the circumstances were clearly rather different, I believed that unless Mburo genuinely engaged its neighbours, when push came to shove the next time, government would lose control completely and all the park and its values would be lost. It was abundantly obvious to me that reaching out to our neighbours, even without sharing power – throwing sweeties over the fence, as it has been perhaps unfairly described – created positive and friendly relations.

Most on the conservation side were happy to stop there. Arthur, Moses and I, though, were sure of the need to go further and to share decisions over what parks were for and how they were to be managed. We also believed that if conservationists and communities could sit down and really talk in a spirit of openness, tolerance would grow into cooperation, then into collaboration and finally into partnership. We might not be able to get there immediately, but we could begin the journey. And that is what we tried to do.

In 1993 things on the governance side picked up speed with the sudden announcement of Park Management Advisory Committees (PMACs). While we had been busy in our small way in our small park, larger projects were at work in the larger parks considered much more important than Mburo. The Development Through Conservation Project – I always felt it would have been more honest and realistic to call it the Conservation Through Development Project – started work in 1988 with funding for the United States Agency for International Development, and implemented by the World Wildlife Fund and CARE International. These were global heavyweights, and they were supporting the recently declared Bwindi Impenetrable National Park, home to half of the world's iconic mountain gorillas.

This confluence of powerful players, amongst which I include the gorillas, meant close attention was being paid by a lot of interested parties. The American government was paying especially close attention. Its agency had been instrumental in creating Bwindi as a national park in 1991, and it became embroiled in the conflicts that arose when people were stopped from using the forest. Not surprisingly, communities were unenthusiastic about this and gathered the support of politicians keen to make capital out of the issue. Encouraged by the Forest Department fighting to retain Bwindi as a forest reserve, and backed by organizations supporting the rights of local and Indigenous Peoples, the opposition to Bwindi Park mounted. This led to a great deal of research and a great deal of consultation, but in the end the global significance of the gorillas and the determination of the directors of Uganda National Parks to get the valuable estate swung the argument. In truth, the deal was always going to go ahead.

Despite concerns to avoid damaging the interests of the people, the law was the law, and only so much could be done to avoid the legal implications of creating the park. To defuse resistance and reduce the danger of real confrontation, as well as to avoid the perception that the international conservation community had made life more difficult for already poor communities, especially the Batwa people,[128] community participation and benefits began to be pushed harder.

Many of us were delighted with the new emphasis, though not necessarily with the speed at which it had been introduced and implemented. Before we knew it, resource access agreements were officially on the table, even in the gorilla parks; the allocation to the communities of a share of the park revenues was proposed, influenced by Kenya's lead; and communities were to have a role advising park managers. Orders were given to establish Park Management Advisory Committees without delay.

Like many good ideas, especially ones imposed from above, when it came face to face with the practical realities it encountered problems. The requirement that each parish adjoining Mburo be represented on the advisory committee meant that it had over 40 members; other parks would have even larger, and more unmanageable, numbers to contend with. Then the advisory committees were to meet regularly, and naturally the park had to pay for the travel, accommodation, and allowances that the representatives not unreasonably expected. The project could cover these costs for the time being – but the long-term sustainability of the committees would become a challenge.

Then, although the existing local government Resistance Committees included responsibility for the environment within their remit, our committees were to be separate bodies. This was because local government was considered too political to engage in park management, and its elected members might expect to share in decision-making rather than simply give advice.

It did not take me long, though, to find out that the advisory committee members were political, too. They were not overly anxious to tell the warden what he should do;

---

128 The Batwa, sometimes called the Forest People, were perhaps the poorest, most powerless and marginalised people in Uganda then and still are today. Originally hunter-gatherers, they had been encouraged, persuaded or pressured to abandon their nomadic lives by British administrators. They were settled at the edges of the forests from which they continued to draw most of their economic and cultural sustenance. The creation of the national parks put an immediate stop to that, and their lives went rapidly from difficult to impossible.

rather, they were keen to tell the people who had elected them what they should do. It became apparent to me that most of the advisory committee members saw it as a rapid route to authority; it was not long before they were asking for uniforms and identity cards, and making plans to patrol their villages.

This was not what I had envisaged. Even though the committees were limited to giving advice, I had looked forward to their members becoming a valuable intermediary on behalf of their communities. But if those members were to become self-styled rangers they could not represent the interests of their communities and would end up representing the interests of the parks. If that happened, they would not even be able to advise the wardens. I had wanted the advisory committees to hold the wardens to account on behalf of their communities, to explain community perspectives, and to carry messages, even ultimatums, from their communities. I did not see the need for a 'friends of the park' group to carry park messages, and I certainly didn't want vigilantes extending the writ of the park.

But the members had their own agendas. Over the next few years the committee at Mburo became less and less useful, and before long its only role was managing the revenue-sharing programme. Even this was denied it when local government reiterated, with increasing confidence, that it was its role to guide local development, which extended to taking decisions over the revenues shared by the park. Once this function was taken from the advisory committees, they had little purpose and withered away.

All of this happened 30 years ago. Since then, the degree to which local communities should share in decision-making in Uganda has by and large turned out to be 'not at all, really, thank you'. Only one of the categories of protected area established by the Wildlife Act of 1996 had provisions for shared decision-making: Community Wildlife Management Areas. Although the managers of national parks, game reserves and forest reserves could work towards sharing decisions with their neighbours and other parties, they do not. The exciting advances made in the 1990s with the resource- and revenue-sharing programmes did not mature into that all-important and fundamentally different decision to share power.

The community conservation agenda of 30 years ago seemed to bode well for a different kind of conservation evolving in Uganda. It seemed that the consequences levied by exclusive protected areas were going to be re-appraised, and true partnerships explored, to encourage and facilitate communities to share in their design and management. Unfortunately, little of this came about. The steps taken during the 1990s were quite radical at the time and supported by many wardens working at the hard edge of conservation. But the advances in engaging communities came about primarily due to pressure from international donors and projects, and people like me. It turned out that real support for these ideas amongst those actually in charge – the directors and board members of Uganda's conservation authorities – was lacking.

Pushing from behind or leading from the front?

Once a year the leading lights of Uganda's conservation community met to discuss the big issues of the time. Unusually, participants were invited to these 'Conservation Forums' as individuals, rather than as representatives of their organizations. The idea was that this would encourage people to speak their mind without having to worry about toeing institutional lines – Chatham House rules prevailed. In 1994 we gathered in the town of Kabale, set in the cool hills of the west. The theme that year was community conservation.

There were so many initiatives moving forwards, so many ideas attracting attention, so many proposals on the table; we wanted to talk over the direction we were all headed in. There were presentations on the revenue-sharing programme and how it might work best; on resource access in the parks and how to manage it;[129] on what community participation meant; on education programmes for schools; and on a host of other initiatives, ideas and proposals. Presentations based on the practical activities being undertaken were followed by lively debates. What was the purpose behind sharing park revenues? Could meaningful participation be achieved without sharing power? Did we want to link or de-link communities and park resources through our interventions? What was working? What was not? What was next?

There was a collegial warmth in the cool air as we contemplated the improving relations with communities promised by these innovations. But then, towards the end of the meeting, an advisor in the Ministry of Finance took the podium. Considering the two days of enthusiastic discussion he had quietly and patiently endured, he was brave. His views were at odds with just about everything that had been said at the forum. He questioned the wisdom of talking to communities. He wondered why there was not greater emphasis on law enforcement. He suggested that the limited financial resources were being wasted. And he bluntly condemned the weakening of the national park ideal.

The rosy atmosphere in the room faded. People around the room looked appalled, and were readying to refute these regressive, outmoded and out-of-touch ideas. Then he concluded with passion: people should be grateful to live next to parks; people should thank their stars their government had the foresight to protect them; people should get down on their knees and give thanks rather than raise their voices with demands and complaints!

---

129 Though perhaps not fully recognised at the time, the establishment of community resource use areas within national parks was a radical and progressive step. It is very unlikely that the national park authorities would take such a decision today.

Before the session chairperson could respond or call on any of the urgently raised hands, the room exploded in applause. Dr Edroma, the national parks director, our leader, rushed to the podium to shake the speaker by the hand and clap him on the shoulder. 'Yes!' I could hear his booming voice above the applause, 'You're right, you're right!'

We project managers and advisors sat in silence, exchanging glances. Some of our closest colleagues were applauding hard too.

The community programmes were too far advanced to be pulled back, and their financial contributions too great to give up, so they continued. Over the years, elements of the community approach supported by our projects became part of the day-to-day operations of the parks, but at that meeting I had discovered an uncomfortable truth; although I'd believed that I and my fellow advisors were supporting our Ugandan counterparts, assisting them to achieve their vision, it turned out that we were actually running out in front, and just assuming our partners were with us. I knew Arthur and Moses were with me, but it turned out that many others were not. Fixed on the future, we did not always look back, and hadn't realised we were alone.

Relations between Ugandan and international conservationists could be complicated. the two parties often seeing things very differently. Too often we, as advisors, acted as though the projects were ours and should reflect our thinking. When these didn't match the interests or expectations of our colleagues, or the demands of the systems they worked within, we were alternately disappointed or dismissive. We understood that we were not in charge, but we sometimes believed we should be, and sometimes acted as if we were. The consequence was that our initiatives too often turned out to be just that – *our* initiatives. When our projects ended and we left our offices, the door was shut and the reports, papers, proposals, frameworks, guidelines, protocols, procedures and who knows what slowly dulled and dimmed with layers of the red dust that finds its way into every corner in Uganda.

This is not to say that all the initiatives developed at Mburo and other parks were forgotten. Many of those initiatives influenced how conservation practice developed in Uganda, and they still provide a foundation on which today's park managers continue to build. The activities implemented at Mburo during the ten years the community project operated did help save the park. Fundamental issues remain, and many of the problems that led to Mburo's fall have not been fully resolved. Enough was done, though, to defuse the worst of the conflicts; enough was achieved to create a degree of collaboration and neighbourliness; and sufficient benefits were shared for communities to feel they were getting something in return for their forbearance.

If you drive to Mburo today, you will enter through a handsome swinging gate and drive easily along well-maintained murram roads. You will see herds of zebras and impalas, pods of warthogs, and topi posing atop termite hills and discover that giraffes were introduced in 2015. At night, from your perch at the bar of your lodge, you may hear a leopard sawing, answered by alarm calls of baboons. But if you listen carefully you may also hear a gentle clop, clop from the lake – not the sound of a night bird but of a fisherman, beating the water to drive fish into his nets. Or you may hear the scrunching sound of long-horned cattle cropping the grasses, pushed into the park on a moonless night to enjoy the rich pastures of Meriti Valley.

Despite the efforts of the community unit, many of the old problems remain. Relations can still be tense, and though the rangers continue to hold power over the park and its resources, communities continue to exercise their *de facto* power to use them, even if clandestinely. But the park has survived – something that did not look likely 30 years ago when Arthur, Moses and I sat down together to wonder what we should do first to make a difference.

5

The Beautiful Land discovered

By 1984 we were nearing completion of our first project. Florence Kabwechera, one of the park's community rangers, arrived outside my tent one morning to remind me that I was to accompany her to spend the day with one of the community groups she worked with. Florence, slight in form but strong and determined, had been one of the first community rangers recruited by Arthur, and she worked under the careful guidance of Moses to deliver the park's community programme.

The Kigarama Hills were barely visible as we headed across the park. An early morning mist turned the air a dirty pink. As we drove we added a heavy pall of red dust which billowed behind, engulfing herds of impala and zebra that raised their own dust as they toiled down to the lake to drink. Grime gathered at the corners of our eyes, and sooty smudges settled and dissolved on our sweaty skin. Meriti Valley had burned the day before and still smoked in places. This was end of the long dry season. We would be hot and dusty before the end of the day, and wishing for rain to settle the dust and cool the air.

The mist lifted as we descended Rwenshebeshe Hill[130]   to cross the swamp to the Sanga Gate, but the view was still hazy and the western hills were reduced to stage flats. A slow guard swung the gate open and we drove up through the banana plantations that increasingly dominated the hills bordering the park, their great ragged leaves coated in the dust thrown up from the road. Florence talked to me about the women she worked with. She had encouraged some to start a small bakery to make sweet bread and buns to sell to their neighbours; some were trading fish from the landing in the park; others were trying their hands at bee-keeping. The Bahima members of the group had settled on handicrafts, using the skills they employed in making *emihaiha* – close-fitting caps woven from grass for covering the polished necks of their milk pots. Someone had found how to

---

130 Place names give the lie to descriptions of an area as wilderness. This hill was named for the wild asparagus that grew profusely on it, a spiny, spindly plant growing to half a metre high. I learned that if I mounded earth on the shoots when they first began to show, I could later harvest a handful of asparagus tips. Though no more than a few centimetres long and little more than a matchstick thick, they were delicious.

flatten these lids to make discs instead of cones. Decorated with traditional patterns and motifs, these would make nice earrings, they thought.

I had been invited as a source of wisdom in general and of urban women's tastes in particular to take a look at these and advise on whether they would find a market. Florence and I perched on a low wooden bench to look at the circles of woven grass, while the ladies, sitting comfortably on papyrus mats, explained how they made them and the meanings of the symbols in their intricate patterns. I had seen similar earrings in the ears of fashionable girls in Kampala and was sure this idea could work.

Bottles of sodas, sweet and warm, were handed around and opened with a gentle hiss. As the morning progressed, the heat building under the thin shade of the acacia trees, we agreed that the project would supply the ladies with the fittings needed to complete the earrings and identify shops in Kampala that would sell them. Within a few weeks the women were delivering handfuls of earrings and we were placing them in shops. They sold well, and orders for more followed. Of all the enterprises Florence was supporting, this one was shaping up best. It could even be represented as one of those illusive win–win activities, as the best grasses for making the earrings were harvested from the park, creating a positive link to income generation.

A few months later I was invited again – the ladies had something new to show me. They had begun to weave delicate chevrons from dyed sisal fibres. Decorated with translations of traditional patterns worked in dark blue, some with tiny beads incorporated into the weave, they looked like feathers dropped by some fabulous bird. They were unlike anything I had seen; very fine, very smart, exquisite and unique. The ladies had used their day-to-day skills and modified customary designs to create something original and special, and they were proud of them. So was I. A stream of these new earrings began to flow to shops in Kampala and the tourist camps in the park, and a not insignificant flow of Uganda shillings came back.

Some months later I was invited to a third meeting. This time, though, it was Florence who had asked me. She had been surprised and disappointed to find that the Bahima ladies had rather abruptly stopped making their earrings. She wanted my help to understand why this enterprise – the one that had been her greatest success, and which seemed to have such strong ownership amongst the women – had suddenly collapsed. We found the ladies waiting to meet us, and though our rickety bench was there under the acacia tree, and the ladies sat on their mats with their legs out as before, there were no busy fingers at work now, no earrings in different stages of completion scattered across the mats. The ladies, not entirely comfortable – they knew that Florence was disappointed with them – tried to explain why they had stopped weaving the earrings.

The story was not easy to follow, as is often the case in such matters. When the talking was done it seemed, in conclusion, that the ladies had simply lost interest in the whole thing. Making the earrings took quite a bit of time, they explained – I had to agree, as it would have taken me an age – and though they were getting a good price for them and had no other way to earn that kind of money, they were just not that fussed about the whole earring thing any more. They happily agreed that making the earrings paid them better than any other activity they could do, but the fact was that they had other things to do. Making money was not such a big deal for them. They were more concerned with making sure they were keeping up with their duties at home, and had begun to feel they were failing in their responsibilities.

The ladies had a clear picture of what they needed to do for their families. Their tasks took up a lot of their time and occupied a lot of their thoughts. They had to make decisions about the milk the cows produced each morning and evening; it was their job to ensure that everyone got their share. They had to plan for the excess milk, which was made into buttermilk or churned to make ghee or sold. Making the right decisions about the milk took more time than one might think, and to get it right they needed to know in detail what was going on with the herds and in the household.

That was not in itself a problem – they had not and never would have compromised on those responsibilities – but the milk was not all that preoccupied them. They also had to ensure that all was beautiful at home. Their houses needed to be decorated, inside and out, and kept clean and scented. Their milk pots and gourds had to be polished and displayed with their covers in place on the platform centre stage in every home for visitors to see and admire. And they needed to spend time on themselves so that they too were beautiful, which also meant they needed time to rest and relax.

I was baffled. These were women from poor households. They might not be the poorest of the poor, and indeed Bahima households were considered rich by farming households; but I had been in their houses and they were as basic as any that I visited. If a *muhima* was in real and urgent need of money to pay school fees or hospital charges, they could sell a cow, but the need would have to be great indeed for such a desperate measure. Day to day, these women were as short of cash as the wives of the farmers. Nonetheless, when it came down to it, faced as they saw it with the choice of earning good money or meeting their own, their husbands' and their neighbours' expectations, it was their responsibilities and their duties that prevailed.

In due course I would learn more about the Bahima households that would help make sense of this, but at the time I could only sympathize with Florence, who was a little put out by it all. It struck me, though, that much of what engrossed the Bahima women seemed related to making things beautiful, and that they had chosen to pursue beauty over the earning of money. Apparently, the lever of financial inducements that we had been hauling on to influence how people thought about the park, was – for these women and their families at least – *not* the right lever. This little earring enterprise had been strong, developed by the women themselves, based on their existing skills, producing something that was selling well and generating good income. It was the most successful of our micro-enterprise SCIP projects. Yet despite all this, its originators and beneficiaries had walked away from it.

Reflecting on this, the event seemed to have something to teach me, if I could put aside the small and nagging concern of how I would explain it in my next monthly report. I realized I needed to think more deeply about poverty: what it might be for the poor themselves rather than what I thought about it. How did people I labelled as poor think about their state – and, for that matter, about that label?

I also needed to consider how my colleagues and I responded to and behaved towards the people we defined as poor. I had believed that a decent life required more than the meeting of basic material needs – but our work with communities seemed to contradict this. Human needs for security, social coherence, identity and self-esteem might be acknowledged at some level, but they didn't influence the activities that the donors funded and the ones we implemented.[131] Given that basic needs for food and shelter, health care and education were not being reliably met, it is perhaps no surprise that more rarefied needs were not raised by communities during our meetings with them.

Unfortunately, the assessments that were an unfailing part of our process of engaging with communities were not designed to probe for the less immediate and less obvious needs. Neither was it particularly surprising that they were not considered, let alone addressed, by our project, given our focus on economic solutions to economic problems. Yet this failure to discuss the things that would turn bare existence into lives of value and meaning excluded whole swathes of our neighbours' lives from our understanding and consideration. In terms of achieving conservation, this meant that many of the ways nature contributed to our neighbours went unrecognized, and the synergies and partnerships that we could have built by recognizing their significance and working to validate these values went unexplored.

Little though the Bahima ladies had in my eyes, I came to realize that they considered their material needs met sufficiently to allow them to walk away from their income-generating enterprise. They had made it clear to me that they were more concerned with their roles within their communities and families than in having money to tuck away.

---

131 Abraham Maslow's paper written in 1943, '*A Theory of Human Motivation*', continues to influence understandings of human needs, in part because it is so lucid. The theory is most often represented by a pyramid in which physiological needs, for water and food for example, form the broad base of the bottom, while more rarefied needs, for esteem, for example, the need to be respected, and self-actualization, the need for one's abilities and potential to be fulfilled – which Maslow believed were shown in the desire to do good and contribute to the greater good – are represented as bands higher up the pyramid. This suggests, perhaps inadvertently, that lower-level needs are more important and stronger drivers of behaviour than higher-level needs, and may explain why higher needs feature so little in development projects, which focus almost entirely on meeting physical needs. Though Maslow recognizes that lower-level needs must be met first, once they have been met, humans require higher-level needs to be met also. Poverty alleviation rightly occupies the attention of strategists and policy makers, but their interventions remain almost exclusively attached to alleviating immediate and physical needs despite the countervailing evidence of the critical nature of long-term and cultural perspectives, and the importance of how people and communities understand themselves and their needs.
'*A Theory of Human Motivation*', Maslow, A. (1943) *Psychological Review*, Vol. 50(4), 370–396.
For example, the Shuar people of Ecuador explained their rejection of mining within their territories in an open letter, writing, 'do not tell us that mining will get us out of poverty, because we, with our way of life, do not feel poor; tell us, instead: how will you protect us as a People and as a distinct culture?'
From the Cordillera del Cóndor 04-01-2017: https://iccaconsortium.wordpress.com/2017/01/25/posicionamiento-del-pueblo-shuar-arutam-24-01-2017

It is easy to slip into the habit of defining people by their poverty (or otherwise) and assuming that this is how they define themselves. It is as easy, too, to assume that the only concern of poor people is to be less poor. Our micro-development projects were all conceived on this understanding. We didn't consider that the values and beliefs that define people and organize communities are powerful governors of behaviour. Even in the most extreme situations of deprivation, however hard the times, social order tends to persist, and communities do not descend into chaos. They are held together by the conceptions their members have of themselves and their people, and the kinds of behaviour acceptable to them, their leaders and their ancestors.[132]

Bahima men carry out most of the outdoor work required by the pastoral lifestyle, tending and watering the herds, milking the cows, constructing the houses and enclosures and collecting firewood. Bahima wives have other responsibilities; they are at once the homemaker and the primary ornament of their homes, which they grace with their beauty and poise. They control the distribution of the household's milk – its day-to-day wealth – employing it to maximize the wellbeing of the household and the herd, and they employ their sexual favours to the same ends.[133] When the earring business interfered with these roles, the ladies abandoned it without hesitation.

Questions about what motivated the Bahima community and what they valued were kicked to the front of my mind by these ladies. While Florence had needed to understand what had happened to her women's group, I needed to understand another mystery.

When we began working to improve relations between communities and the park, I had expected that the most difficult people to convince of the park's relevance would be the farmers, and that the pastoralists would be less fiercely opposed. My thinking was

---

132 The plight of the Ik people of northeastern Uganda when excluded from the Kidepo Valley when it was made a national park in 1962 is referred to in Chapter 3. Colin Turnbull's *The Mountain People* describes what happened, at least as he saw it. One of the reasons the book is so shocking is that it describes the Ik, so ground down by poverty and hardship, losing their basic humanity. Pushed from their farms and hunting grounds, a three-year drought plunged them into extreme hardship. Turnbull wrote that they no longer cared for their old, who were left to starve in their houses, or for their children, who were left to die in the bush. Individuals cared only for themselves. Turnbull proposed that the government of Uganda should disperse the Ik community as individuals and small groups across the country so that their failed morality and sense of community would not infect the wider society.

The book has been much criticized since its publication, and other visitors to the Ik (who survived the sixties drought and the Karamoja famine of 1980/81), including myself, have observed none of the behaviours described by Turnbull.

133 As in many traditional societies, sex outside of marriage was anathema to the Bahima, and in some cases punishable by death. Married women, however, had great licence in sexual matters. A married woman was not expected to be faithful to her husband, and used sexual favours to create relationships that would benefit the family and the herd. The spear of a stranger planted at the door of the eka was a sign for the husband to keep out.

based on what seemed to me to be the clearest conflicts relating to each of these groups with the park.

The most obvious interest for the farmers in the park was the land itself. The farmers wanted to occupy the park, and farm its ample and fertile valleys and hills.[134] There was no possibility of this being agreed, and it would not even have been a subject of discussion. Indeed, Arthur, supported by the project, was working hard to find ways to remove the farmers from the park.

The next issue of greatest concern and interest for the farmers was the wildlife. On the one hand they wanted to hunt it and on the other they wanted to protect their crops from it – a visit from a family of bushpigs would turn a planted field into a ploughed field overnight. Hunting was also off the negotiating table, and even allowing farmers to shoot any animals causing problems on their lands could not be agreed unless they were classified as 'vermin' by law. Moses instead helped farmers establish wildlife-proof hedges and a wildlife control unit manned by rangers, who would do any shooting that might be required.

The Bahima, on the other hand, had no crops to be damaged and no interest in hunting. They wanted access to pasture and water, and the park had agreed that they could bring their herds in during droughts, reinvesting Mburo with its historical role as a grazing and watering resource of last resort. I had imagined that the Bahima would at some level recognize that the park was preserving Mburo as an example of their cultural landscape – open grasslands free of fences, farms and bananas – and support it accordingly; the park was thus, as I saw it, keeping the land pretty much as the Bahima would have kept it themselves in a different world. I recognized the obvious area of difference – the barring of their cattle from the land – but felt it would be easier nonetheless to build bridges with the pastoralists than with the farmers.

But a few years of working to influence community relations by supporting SCIPs, holding meetings, having discussions, negotiating agreements, sharing revenues, working with schools, and all the rest demonstrated that my assumptions had been quite wrong. While our relations with farmers improved, our relations with the Bahima did not.[135] Could this be simply because the pastoralists were less interested than the farmers in what we had to offer them? The project had been designed to deliver financial and material benefits, but perhaps these were insufficiently attractive when compared to what the park denied them.

---

134 The agricultural peoples of southern Uganda have a settlement pattern based on families living on their land rather than living in what we would call a village – houses gathered together in a central location. But a Banyankole farmer will, along with his family and workers, live in the middle of, and surrounded by, his land. Then an area of farmed and fallow fields scattered with dwellings, bound together as a community by traditional institutions of leadership, is called *ekyaroa*. It is translated into English as 'village' – but in reality it is a very different entity.

135 Information collected by interviewing people living around the park at the start of the project and after five years of implementation found that the Bahima started with, and retained, significantly less positive views of the park than did the farmers.
    'Community attitudes and behaviour towards conservation: an assessment of a community conservation programme around Lake Mburo National Park', Infield, M. and Namara, A. (2001) *Oryx*, 35, 1, pp. 48–60.

Watching the magnificent herds of Ankole cattle sailing like flotillas of yachts across rippling seas of grass, and observing the avid gaze of their owners as the animals dipped their heads to drink, listening to stories praising legendary beasts of the past or the heroes who fought to protect them, it began to dawn on me just how important the cattle were to the Bahima – not just in relation to an individual and his animals but, more profoundly, to the idea of the cattle's relevance to the Bahima as a people.

I began to feel there was something that underlay the arguments made to justify the demands to access water and grazing in the park, something mysterious and obscure. When lobbying the authorities, the pastoralists expressed their position in economic terms; if the milk reduced their families would suffer; if the animals died school fees would be missing. But as often as I heard these arguments during the sometimes-heated meetings with park staff, the passion that the Bahima had for their cattle seemed strangely absent. I had myself seen cattle being driven into the park to graze when there was more and better grass outside. This seemed to refute the simple economic arguments that both sides insisted on as the cause of, and the solution to, the problem of the cattle and the park. Something else had to be driving behaviours and practices that otherwise didn't make sense – something that was not being expressed or acknowledged.

Once I began to suspect the existence of a hidden truth behind the positions projected by the Bahima, I could not help wondering at the vehemence of my colleagues' rejection of any possibility of the presence of cattle in the park too. They couched their arguments in terms of protecting biodiversity, in itself a complex proposition; or they talked about the need to reserve grazing for wild animals; or they raised concerns that tourists would stop coming if they saw cattle when they had paid to see wildlife. If these problems were really what drove their insistence on excluding the cattle, there were surely ways to deal with them other than a blanket ban on the animals. Might there be other explanations for the staff of Uganda National Parks being so uncompromising – explanations they were hiding from me, and perhaps from themselves too?

My suspicions that the positions taken and defended so fiercely by both the Bahima and my colleagues derived from concealed or unrecognised influences now merged with my own misgivings about the nature of exclusive national parks. If my insights into what motivated the Bahima could be so wrong when it came to the relatively simple matter of their livelihoods, how much more wrong might I be when it came to their relationships with the natural world? After all, the influence of the material world on behaviour had to be a simpler concept to grasp than trying to ascertain how feelings influenced behaviour towards nature. After all, did I even understand my own feelings for nature and their influence on my behaviour? If I could barely comprehend my own emotions, they would surely be a poor guide to understanding those of the Bahima, or indeed anyone else's.

These ponderings might seem rather abstract, but I had a compelling and practical reason for delving into them. Our community approach at Mburo was not making headway, and I began to feel that this was the case more widely; despite the best efforts of conservationists, nature was clearly in retreat not only in Uganda but around the world.

There was no question in my mind about the need to work with our neighbours locally to help reverse this, but perhaps the assumptions we made about what interested and motivated people, what lay behind their relations with and behaviours towards nature, were leading us in a wrong direction. Perhaps, without acknowledging it, the way we experienced nature at a personal level, secreted behind the edifices of science and rationality, was affecting the way we went about trying to conserve it. Perhaps it was a lack of openness and honesty with ourselves that was causing errors in the way we engaged with local communities.

After prevarication, uncertainty and argument back and forth with friends and colleagues, I put an idea to Dr Eric Edroma, who had done so much to support our work at Mburo. I had first met him in 1981 during those heady days of the aerial survey when my excitement at being in the wilds of Africa was fresh enough to suppress the misgivings I was beginning to feel. He was then the director of the Uganda Institute of Ecology at Queen Elizabeth National Park. We would chat in his small, dusty office about the decline of science in the parks and eat fruit salad laced with waragi firewater.

By 1995 he was in charge of Uganda's national parks. Despite his apparently instinctive rebuttal of community approaches, revealed at that awful meeting in Kabale, he agreed to my proposal.

Arthur, now head of the new Community Conservation Department, was full of encouragement for my proposal, too. The second phase of our project was about working to institutionalize the lessons from Mburo and elsewhere, and he was keen to explore community engagement from all angles.

My proposal was to investigate the true nature of the deadlocked conflict at Mburo. I wanted to understand why the Bahima continued to reject the park despite all our efforts to provide them with benefits from it. I wanted an explanation for why they pushed their cattle into the park even when there was water and grazing outside, and even at the risk of fines and beatings from the rangers. At the same time, I wanted to know why many of my colleagues were so adamant that the cattle had to be kept out at all costs if Mburo was to be protected.

It troubled me that this was still the default position, even after all of Mburo's troubled and shambolic history and the practical day-to-day challenges of endlessly excluding the Bahima and their herds. If I could appreciate the implacable rejection of each by the other, and comprehend the reasons underlying their arguments, perhaps I could help find a way for the park to engage with the Bahima and win their support. This would surely strengthen Mburo and its values. It might even elucidate the pure pastoral values that had led in the past to the area of the park being selected for the king's herds, values which the Bahima clearly still cleaved to, despite their opposition to the park itself.

I had not initially thought in terms of formal academic research, but I feared that any suggestions coming from me (considered something of an irritant in some quarters) about changing relations with the Bahima and their cattle would be rejected without stronger arguments. Rigorous research through a respected university would help my ideas get a hearing. Anything less, and my colleagues might simply dismiss them as just another of my notions.

It took a while for these ideas to develop in a practical form, but on a chilly October afternoon in 1996 I found myself sitting in front of my academic supervisors in the University of East Anglia. A few months earlier I had sent a short note to a Professor Stocking, and he had written straight back agreeing to take me on as a PhD student. I had expected some discussion, some exchange of letters – but before I had fully engaged with the possibility, I found I was going to be a student again, dividing my time between managing our Phase II project, advising Arthur and supporting the Community Conservation Unit, and undertaking research.

'So, Mark, tell us about your idea.' Professor Stocking was peering at me over his glasses.

I threw myself into the story of Mburo, its history, the Bahima, their extraordinary cattle, and the conflict that wouldn't go away.

When I had finished, no doubt with a satisfied flourish, he asked again, encouragingly, 'Good. Good. But what is your idea here? What is your research going to be about?'

After a pause during which my mind spun, I narrated again, in greater detail, the context of the park, the people and the conflict, until gently interrupted.

'But what is the idea in all of this, hmm?' he asked.

I was nonplussed. I worked in conservation. I was a project manager. I wasn't sure I had ideas; I had observations and wrote reports. I floundered on for a minute or two.

'Tell us,' he interrupted, still encouragingly, still peering over his glasses, 'Tell us what your *research* is about' – a pause – 'without talking about the park, these Bahima, or their magnificent beasts.'

I was stumped. I tried to explain that I wanted to understand what was going wrong at Mburo so we could fix things, do things differently, get better. Patiently but firmly he explained that that was a *project*, not research. I would need to be able to tell the story in general terms, and if I wanted to do that I would need to think more deeply. He advised me to spend a few weeks in the library, do some reading, have a think, and come back for another chat. I was out on the grey concrete walkway almost before I knew it, walking, a little baffled but doggedly, towards the library.

I don't think I had read anything, thought deeply, or had anything that could be described as an idea about nature and conservation, for years. The pell-mell world of projects and interventions and reports seemed to leave little time for such apparently rarefied pursuits. My few weeks in the library were a revelation. I discovered that almost everything I had learned about ecology as an undergraduate had been replaced by a more fluid interpretation of the natural world. I found discussions about nature and its conservation were going on amongst philosophers, psychologists, geographers and economists as well as amongst the new breed of ecologists.

Entirely new fields had come into existence. As an undergraduate I had been able to choose between zoology, botany or biology: now I had to wrap my mind around conservation biology, environmental technology, political ecology, environmental economics and many more perplexing and obscure subjects and disciplines.

Over the five years of my research I learned to tell my story about the cultural underpinnings of conservation conflicts without talking about Mburo, or the Bahima or their cattle. With great patience, my supervisors helped me articulate the Mburo story in a way that was robust enough from an academic perspective and practical enough to help sustain Mburo. They helped me develop not so much a hypothesis as a purpose for my thesis, which was to examine the nature of national parks as institutions and how the cultural values of those with interests in them affected perceptions and response to them.

Importantly, I found I could begin to describe what I learned from Mburo in terms that could be applied to other parks, to other situations, and even to the conceptions that were driving modern conservation. The ability to generalize from the case of Mburo to wider questions about the conservation endeavour was my reward for my tussle with the demands of academia. It changed the work I would do in Mburo, and how I would look at the entire conservation undertaking.

A couple of years and many days spent in the field after that first meeting in my supervisor's office found me sitting once again in the scant shade of a thorn tree as the dust stirred by the just-departed herds of longhorns settled. I was half-listening to the older owners and their younger helpers talking over the pastoral affairs of the day – the health of the animals, the quality of water in the dam, whether the pasture would last until the rains came.

I had spent the morning in the dim recesses of the house of a recently settled muhima on his newly allocated plot. A number of men, all of whom had been allocated land under

Left: A Bahima home on Ranch 10, where pastoralists
were allocated land to adopt a settled existence

Right: Patrick Rubagyema undertaking field research on Ranch 10

the Ranch Restructuring Scheme, had gathered to generously answer yet more of my questions. When I had run out of questions and they patience, we set out to find the herds and follow them to the afternoon watering. We ambled through the bush behind the quietly grazing animals as they skirted thickets and stands of acacias, descending finally to the valley where a small dam had been dug.

We were on the portion of Ranch Ten, which under the 1960s Ankole Ranching Scheme had been divided into plots and distributed to landless pastoralists. This initiative, which the Community Conservation project had supported, had been designed not only to end the conflicts between the Bahima, the ranchers and the park, but also to end the nomadism that was considered both impractical and inappropriate at a time when everyone else was securing land for their individual use. Nomadism was seen as a sign of backwardness.[136]

I had been spending a lot of time on Ranch Ten. Chabyhindi, one of those recently settled pastoralists, had generously accepted me as a member of his eka, and had encouraged me to construct a hut inside his compound. I was able to spend time there whenever I wanted, and it became the base from which I began my efforts to learn about the particular cattle culture of the Bahima. With notebook in hand and my indispensable guides, cultural interpreters and language translators, first Charles and then Patrick, at my side I did my best to adopt the guise and methods of a social anthropologist. I

136 The president felt so strongly about the ills of nomadism that he frequently berated his people – the president
a *muhima* himself – for being backward. He even created the position of Minister of State in Charge of Ranch
Restructuring, Water Development and Anti Nomadism.
Custodians of the Commons, Pastoral Land Tenure in Eastern and Western Uganda, Kisamba-Mugerwa, W.
(2014) ed. Lane, C.R. Oxford: Earthscan, p. 232.

Troughs and dams for watering the cattle, constructed from clay are located on the edge of the wetlands; they are filled by hand

engaged strangers in conversation as they walked through the bush; I wandered through homesteads, poking my head into stalls and sheds; I talked to children and asked them about the games they were playing; I quizzed women churning milk to make ghee; I drank mug after mug of sweet milky tea. I took notes, kept a diary and worked at my 'participant observation' – the classic method of understanding by doing.

I had been watering the cattle. A broken jerrycan had served as a bucket to scoop up water from between my legs. Bare feet sinking into the mud, I swung the bucket up, balancing myself against its pull as it broke from the surface of the water, suddenly heavy and unmanageable. Controlling its momentum and arc carefully, I settled it onto the friable rim of the mud trough, trying not to damage it, and cautiously poured the water in amongst the jostling heads and horns, removing bits of grass and leaves mixed into the swoosh of water as I had watched my mentors do.

Pouring water into a trough does not sound difficult, but if I startled the animals I would get a stern rebuke; if I cracked the soft edge of the trough there would be a groan or a giggle from my watchers; and if I slipped or stumbled in the sucking mud, there was just silence and raised eyebrows. I was of course too old to be lifting water. This was really a young man's job. But as a man without cattle of my own, I was considered poor in social as well as material terms, so it was not entirely inappropriate for me to perform this menial task.

Having a white man spend time in the community was one thing, but letting him undertake tasks he was unsuited to and which might affect the cattle was another. The owners of the herds that came jumbling down in turn from the hills watched me with a mixture of amused indulgence and concern, and I knew that pointed questions had been asked of Chabyhindi, my patron.

My time was spent trying to gain some comprehension of the meaning of the evidently special relationship between Bahima, the long-horned cattle and the land. My 'how to be an anthropologist' book[137] sat with my notebooks in my decorated, thatched, but brick-built hut[138] Despite the months I had been visiting, not to mention the years I had been living in Mburo, my command of the Runyankole language was virtually non-existent, going little further than the wide range of greetings used[139] So I depended on my questions and their answers being translated for me, which could be exhausting for all parties.

The heat of the day had reduced, but the sun was still hot as it angled through the trees, and I was sleepy from my exertions with the watering. With only a vague grasp of what the herders were talking about, and finding it hard to concentrate, I had drifted off in mental excursions unrelated to where I was and what I was supposed to be doing. Patrick, tall, gentle, thoughtful, and deeply committed to Bahima values, had joined the men sitting on the raised bank of the dam to watch the cattle grazing around us. The rhythmic crunch and crush of grass grinding between bovine molars was mesmerizing. Soon they would begin the slow trek back to their eka for the evening milking.

Blue-spotted wood doves were making their soft, mournful songs of descending notes from the thickets, and chinking alarm calls rose from blacksmith plovers disturbed by the cattle. A low murmur of voices emanated from the group as hands were raised in elegant gestures, punctuating key points. The emphasis of a recurring word, like a dripping tap, gradually penetrated my sleepy mind. There it is again. *Enyembe* something. Is that it? I've heard it before, perhaps … yes, certainly. What does it mean? It seems important. Was that it again? What are they talking about? The cattle of course, always the cattle. It must be to do with the cattle, something I really ought to know by now, I'm sure.

The discussion broke up and we began to trail off to our different homes. I fell into step with Patrick as we went. What, I asked, was that word the men were using every second word? Which one? The one I kept hearing. The one everyone seemed to be using. Something like *yemiwa*, was it? Or *yembiwa* or *embiwa*?

After some puzzling Patrick confirmed with some surprise, 'You mean '*enyemibwa*'. But surely you know that word.'

As he explained its meaning, its significance washed across me and my thoughts began

---

137 Russell Bernard's *Research Methods in Anthropology* was a life-saver for me, presenting a vast amount of information, engrossing examples of field studies, and practical guidance on how to go about being a field researcher. I called it my 'Big Blue How to be an Anthropologist Book', and rarely went anywhere without it: *Research Methods in Anthropology: Qualitative and Quantitative Approaches*. Bernard, H.R. (1995) London: Alta Mira Press.

138 Chabyhindi and I had talked through the design of the hut I was to live in when I was on the ranch. I explained the idea behind the kind of research I was doing, and that it would be strengthened the more I followed the local ways. He pointed out that now he and his community were being asked to settle permanently on the pieces of land allocated to them, they needed to start changing their ways, and changing from temporary wooden structures to more permanent brick structures was going to be part of this.

139 I am spectacularly bad at languages, but took solace from Sir Richard Burton, the explorer. Although he is reputed to have been fluent in 40 languages and dialects, he found the languages of East Africa impenetrable.

to race. I had to stop walking to absorb it. I leaned on my pastoralists' cane as a rush of excitement left me unable to talk for a moment.

Still slightly bemused at my persistent ignorance after all these months and all my professions of interest – how could I *not* know about this most basic of Bahima concepts? – Patrick began to explain the significance of 'Enyemibwa'.

Bahima are very close to their cattle; a young Muhima contemplates the beauty of his bull, Ruhuga

'It means "a beautiful beast" ', he told me.

'Well, aren't all Ankole cattle beautiful?' I asked.

'Not at all! Obviously not. Just look at them. Look at that one,' he commanded, raising his stick to point to a young bull thrashing his horns in a thicket. 'That one is clearly not enyemibwa, not beautiful. Look at its colour. It is a pale brown, not a rich brown. Its horns are big and have a good shape but they have black tips. Horrible! Now look at that one over there. It is almost perfect, a Beautiful Beast, for certain.'

But doesn't everyone think that their own animals are beautiful? I asked. 'They're all Ankole cattle after all.'

'If they are not enyemibwa, no. They would be laughed at. Everyone knows what is and is not enyemibwa.'

Imagine a group of Bahima looking at a herd of cattle. After a little debate, a little discussion, a little emphatic pointing and gesticulation, all would agree which of the animals before them qualified to be called enyemibwa. It was not a matter of opinion – beauty in this case was not in the eye of the beholder – but about applying a set of known, agreed and absolute criteria. In time I was able to pick them out myself. Like judges at a dog show assessing the characteristics of a prize breed, the Bahima check off the requirements of beauty. Does the animal have the correct shape? Is it tall, long-legged and rangy? Does it have a straight back and just a suspicion of a hump nestling a little behind its shoulders? Has it the required deep red-brown colour? And critically, does it carry horns that are elegant, balanced and white to their tips?

Unlike the judges at a dog show, though, they would not be adding scores to find the best animal. To be enyemibwa, it must have all the characteristics, and all of these need to be near-perfect. There might be some debate about whether the horns were truly elegant or sufficiently luminous in their whiteness, but there was no point in discussing whether a black cow, for example, however perfect in every other respect, was enyemibwa – or indeed a cow with small horns. Everyone knew the criteria. If there were just one Beautiful Beast in a herd, everyone would know which one it was.

Perhaps this does not seem so different to the thinking of other lovers of cattle, of which there are many. I am sure that most farmers think their animals are beautiful. I know the manager of the farm where I grew up thought his Sussex cattle were beautiful, and would stop and lean over the gate to point out their powerful dark forms silhouetted against the rough pasture of his fields, or stamping and steaming in the barn on a cold afternoon with the winter light falling on their deep brown hides. A Dinka boy or a Nuer from the cattle-keeping tribes of Sudan, or a Karimajong from eastern Uganda, has from birth his own special ox, dedicated to him, that he loves and praises and decorates. Compared to any other beast he finds it beautiful, he sings songs of love and admiration to it, and when it dies he will mourn it like a family member. Though each muhima loves their animals, and have their favourites, as a people the Bahima have a shared vision of the perfect beast. It is the pursuit of this beauty that defines them as individuals and as a group. Enyemibwa define what it means to be a muhima.

My mind danced around this single word. While 'enyemibwa' may translate simply as 'a beautiful cow', as in 'look at that beautiful cow', the word also defines the meaning and value of the Ankole long-horned cattle to the Bahima. My discovery of this word was a pivotal moment for my research. So much of what I had been told, seen and experienced suddenly came into focus and began to make sense. I started to feel that I could begin to arrange the disparate bits of information I had gathered into a pattern and might finally approach a solution to the puzzles of the earring-making ladies walking away from their business, and the project's failure to engage effectively with the Bahima.

'Enyemibwa' would give direction to my research and help reveal how Bahima relations to their cattle and their land determined their relations with the national park and the kind of conservation it represented. As my analysis proceeded and deepened, my perspective of my own efforts to conserve nature began to change too – and then I found it disconcerting that this single word, this apparently simple idea, could change so much of my thinking.

I began to appreciate why the ladies had stopped their earring-making. I would have to think again about how we should engage with the Bahima community. 'Enyemibwa' helped me understand why I was so uncomfortable about our education programme, despite the near-universal praise for it from community leaders. In time it would help me question not just whether our interventions were appropriate, but what our intentions actually *were*. This single word would lead me to question much that I had taken for granted about conservation and my role in it.

Put simply, the very concept of Enyemibwa, Beautiful Beasts, made me realize that telling the people what I cared about, and demanding that what was important to me should be important to them too, was neither right nor likely to be effective. Clearly, I had to reverse this thinking. What people thought about their world and what was important to them in it should become the centre of conservation initiatives. What was important to me would always remain important to me, and I was confident that my values would be expressed. But my own perspectives needed to take a back seat while our neighbours did the driving.

In the months that followed I learned how people described and identified their Enyemibwa, why they valued them, and how they bred them. I was never able, though, to discover why the specific characteristics demanded were so desirable. Enyemibwa did not produce more milk than others, they did not have a better temperament, or walk further or faster or for longer without needing water. People told me just that they had inherited the long-horned cattle from their ancestors, that their ancestors had loved them for their beauty, and that was why they loved them too.

The love of the Beautiful Beasts was so profound that it influenced every aspect of a muhima's thinking and being. It affected their aesthetic and informed what they considered beautiful in the things around them. They decorated their houses, their walking sticks and their milk pots to reflect the contrasting creamy white of perfect horns with the rich chestnut brown of Enyemibwa. They went to great lengths to repeat this combination. Sometimes I would find in a corner of someone's home a dish, white or cream, with a scattering of the small black and red seeds of the *Ormosia* bush, like a herd scattered across a valley.

The milk pots used by Bahima are a good example of the lengths taken to represent the essence of Enyemibwa in their day-to-day lives. *Eychanzi* are carved from the soft wood of Albizia trees. When new and untreated they are a light tan. Before any milk is poured into the narrow neck, the pot needs to be prepared. It is rubbed with ghee and ashes, then buried in the earth for some weeks. When extracted, cleaned and polished, the dull brown pot has been transformed to a gleaming rich red brown, the colour of a Beautiful Beast. Now the pot is fit to be a receptacle for the creamy milk that is the colour of the horns of an Enyemibwa.

Children will spend hours peeling the bark from long curving twigs collected from just the right bush to make the white horns of their imaginary herd which they will line up on the ground. Or they will sift through the sack of beans in the kitchen to select the deepest, reddest beans to strew across a dusty skin on the floor to be their herd of Enyemibwa.

Ankole cattle have horns that can be 2.5 metres from tip to tip and so heavy that the animals sometimes seem to struggle to carry them. Something as exceptional and unlikely

A Muhima boy shows off his own Enyemibwa from selected red beans and carefully peeled twigs

as this does not come about by accident. It comes through the exercise of knowledge and skill, pursued with commitment and energy, and driven by a consuming desire. The Bahima are the proud possessors of these extraordinary cattle because they are more important than anything else in their world. Though the people created the Enyemibwa, it is the Enyemibwa that define the people.

It was this, I came to believe, that explained the decades of conflict between the Bahima and those that sought to change them or control them or limit their adherence to a lifestyle centred on their Enyemibwa. This was the nub of what I had set out to examine. The conflict between the Bahima and the managers of Mburo was underlain by a cultural identity wrapped around the love and pursuit of Enyemibwa. By breeding these animals the Bahima honoured their ancestors. This required that the land be given over to the cattle entirely; Kaaro Karungi, the Beautiful Land – gave rise to Enyemibwa, the Beautiful Beasts. The relationship of the Bahima with the land was mediated through their love of their herds of long-horned cattle.

A muhima's herd must cater for the needs of the whole family. Children need milk to grow tall and strong; adult men need energy for the demands of the pastoralist life that sees them ever on the move, ever active; girls must have milk to grow fat and sleek as they prepare to marry; women must have the resources to manage the milk and the family and keep everyone on track in pursuit of the good life. All of this comes from the families' cattle.

Many other pastoralists have also been highly dependent on milk, meat and blood for their nutrition, but the Bahima made an absolute thing of it. Though all around them there were people farming bananas and cassava and sweet potatoes, and raising goats and chickens and catching fish, Bahima would touch none of it. Bahima drank milk in the mornings, buttermilk while out with the herds or about their business during the day, and milk in the evenings, and they ate meat and drank blood when the management of their herds allowed it or when ritual demanded it. Anything else was taboo. They didn't even hunt the wild animals that shared the range with their herds. Mixing milk with any other food, it was believed, would harm their animals – not the people, but the *cattle*.

A muhima who ate food not derived from their cattle had to be purged, ritually purified, before they could take milk again. The exception to this rule was millet beer. Everyone needs to relax occasionally, and as good fortune would have it, millet beer, pale and creamy, could, at a stretch, be considered to be like milk. Otherwise, the prohibition was taken very seriously. History records only one act of violence against the British as they moved to control the Ankole Kingdom: the murder of Henry St George Galt in 1905. It seems that the dismay felt by the Bahima when they learned that their exiled king had been forced to eat eggs and other foods had led to the killing. The unease about the consequences of the king's pollution had led to anger that was turned against Galt, who is reported to have ridiculed the Bahima for their dietary restrictions.

Dependence on their cattle for their own subsistence, made more acute by their rejection of all other sources of nutrition should, it would seem, have ensured that the Bahima would work hard to maximize the productivity of their cows. But apart from a young married man, just starting out, who would not have the luxury of thinking too much about the beauty of his herd, this turns out not to be the case. Ankole cows don't give much milk – not much more than five pints each a day in good conditions, and as little as two when conditions are poor. Though the milk of Ankole cows is rich compared to that of other breeds, if you are consuming nothing but milk you need to quaff quite a few pints every day to stay fit and strong. At around 300 calories a pint, the men of the family will need to down several pints a day. And if you are a young woman, wanting to put on weight as your wedding day approaches, you will need to be drinking a good deal more than that.[140]

To feed a family of six during the dry season takes around 20 milking cows, but not all the cows in the herd will be giving milk at the same time. Some will be too young, and some will be pregnant. Some will be getting old and producing less milk, while some will be feeding their calves, and the milk will need to be shared to ensure the calf can grow. And about half the herd will be male.

So a muhima needs to accumulate about 80 animals before he can feel some degree of assurance that he can meet the basic demands of his family. There are many ways he can do this. He will certainly be given some cattle when he marries, and he can ask for gifts from family members or close friends on different occasions. He may be asked to look after some animals if their owner wants to split his herd to spread the risk of disease or drought or raiding. This means he can take the milk as his own, and over time he will keep some of the calves to swell his own herd. And he may be able to borrow some animals.

While busy with this herd-building, a social process that establishes and strengthens connections with a network of relations and friends, all the while making sure his herd receives the best care and attention, a muhima doesn't have the leeway to focus on the beauty of his cattle. When asking for a gift, he might lust after a creature with magnificent sweeping horns, but he will settle for the one with short horns as it will give as much milk as the other. Once the needs of his family are met, though, he can finally begin his life's true work, building his reputation as a person that breeds Enyemibwa, who owns Enyemibwa and who knows how to care for them. This is what will define his place and

---

140 Though Bahima men are generally thin as sticks, Bahima women want to be the opposite. Just as Bahima have an idealised vision of a Beautiful Beast, they have an equally prescriptive view of what constitutes beauty in a man or woman. Both should be tall and elegant, have unblemished dark brown complexions, and have white teeth and black gums. In addition, a muhima woman should be fat. John Hannington Speke found himself in trouble in Victorian England for describing how he had measured the waists of the wives of the King of Karagwe, a long-horned cattle-keeping people he had spent time with on his journey around Lake Victoria. Though their girth had been pointed out to him with pride by the king, the Victorian public felt it unseemly to measure them and then to write about it.

reputation in society. Numbers is not enough; the animals must be Enyemibwa, and they must be cared for.[141]

Now that I believed I had some idea of what was going on, as a researcher I needed the data to test it. So I designed a series of exercises. To figure out the particular characteristics that the Bahima favoured in their cattle I asked owners to point out the animals they considered to be their best. For each one they picked, I asked Why? What made it so good? What was it that made this animal special? By the time I had finished, I had 700 descriptions of the best animals in Ankole. In almost every case they were picked because they carried the specific characteristics of beauty.

It was the loveliness of this beast's colour, neither too dark – black is not considered beautiful – nor too light, but just the perfect *bihogo* colour, a rich red-brown that shines and glows like polished mahogany. *This* cow had a scattering of white stars across her hindquarters, while *that* bullock had a blaze between his eyes; the shapes, locations and the size of white markings could add or detract from an animal's beauty. *This* cow was tall and elegant. Most often an owner would point to an animal's lyre-shaped horns and ask me if I could see how graceful they were, how perfectly white, how fresh and pure they looked. Almost none of the owners selected cows because they gave a lot of milk, or had a good temperament, or put on weight fast. There was no doubting what excited these owners, and it was bovine beauty, how closely their animals approached being enyemibwa.

It was clear the owners picked out the animals for their appearance. If this was also the basis on which they selected beasts for breeding, it would prove that Bahima were actively pursuing beauty rather than productivity. Though an animal had to be near-perfect to be considered enyemibwa, some characteristics were nonetheless considered more important than others. One of these was size; an Enyemibwa needed to be tall. Perhaps this meant that although the Bahima admired beauty in their cattle, they were actually breeding them to be big and therefore productive.

I designed a game in which pastoralists could choose between favoured characteristics. I prepared paired images of animals that were either large or small, had or did not have perfect horns, and were or were not bihogo in colour. My game prevented all three preferred characteristics being selected at once. If the player couldn't have all three in one animal, which would they pick? It turned out that a beast that combined beautiful horns and the right colour was always chosen first, even though this meant it would be small. If this combination was not available, then people would choose a large one. These were preferred historically because they could walk long distances to water. Now it was because

---

141 President Museveni is reputed to own 4,000 Beautiful Beasts and to know the name, history and origin of each one.

they sold for more money. But small animals, like those kept by other tribes, were just not beautiful and could never be enyemibwa.

For any farmer, traditional or modern, selecting breeding stock is the key to getting the offspring they want. Over time, and motivated by a different set of values, the Bahima could have bred more productive cows, but that was not their intention. Indeed, going back to the start, the Bahima – or, more accurately, their Bachwezi ancestors – had bucked that trend.

Ankole cattle are generally believed to be descended from the first cattle to arrive in Africa, *Bos taurus taurus*, that were crossed with the next kind to arrive, *Bos taurus indicus*. Both are descended from the wild aurochs, *Bos primigenius*, but are distinct from each other. The taurine breeds are described as humpless, while the indicine breeds carry prominent humps and are called *zebu*, or humped.

When cattle first arrived in Africa and where they came from is disputed. Some argue Africa's cattle were domesticated in North Africa. Most, however, believe that long-horned humpless cattle first appeared in Egypt some 7,000 years ago, coming from Persia. Whatever their origin, this new addition to the human household was taken up enthusiastically across the continent. Some 3,500 years later, animals with humps as well as long horns arrived, also from Persia. In Ethiopia these two breeds were mixed to create breeds with long horns and small humps, known as sangas. The long-horned Ankole cattle are sangas, as are others in the region, including the famous Royal Inyambo cattle of Rwanda. A thousand years after the magnificent sanga breeds had been developed, they were replaced by the smaller, hardier breeds with large humps and short horns that had begun to arrive from India. These zebu cattle were easy to keep and more productive in harsh conditions, and they quickly swept the board. Only a few tribes hung on to their humpless cattle or sangas, accepting a lower level of production in return for their other values.

The Bachwezi did not made the change, and neither would the Bahima. They refused to replace their elegant, humpless, long-horned beasts with these diminutive creatures with their wobbly humps, almost hornless and lacking grandeur, even if they did give more milk.

The Bachwezi had been no mere mortals concerned with the size of a cow's udders. They were demi-gods after all. They sang praises to their animals, composed poems to them, and traced their ancestry through the generations. They serenaded them with flutes as they walked the hills and valleys. Pacing with long strides, they followed their animals as they moved with swaying horns into the wind, smelling the sweetness of fresh rain and green pastures in the distance. They were unencumbered by homes or possessions, and unrestricted by fields or fences. The world opened to them and their cattle.

This was the tradition the Bahima inherited. They would swallow with pride the costs levied by breeding for beauty, as their ancestors had. This was why they had no time for those mealy-mouthed white men who came demanding that they produce ghee to trade, or worse, insisting that they sell their animals. This was why they resisted farming, fencing or any idea that resulted in enclosing their beasts.

The desire for production that motivated the British, the farmers and even the new Bahima élite, their educated brothers, meant nothing to the Bahima. Their decision had been made long before, when accepting the gift of the cattle from the Bachwezi and the responsibility of their care. Their identity was defined by this burden of honour. Whenever I asked why enyemibwa had to be that colour or carry those horns, I would receive one answer: these were the cattle the Bachwezi had loved, so it was the Bahima's task to keep them just as they were. Rather than submit to the pressure exerted on them by the British to give up their pursuit of beauty, the Bahima would depart their lands, just as the Bachwezi had departed the world.

It takes time and commitment to build a herd of enyemibwa. It takes sacrifice, too. There are other tribes in the region who also keep forms of the Ankole breed, but it is easy to see the differences between their herds and the herds of the Bahima. The Bahima favour bihogo; the horns of their animals are much larger than those of their neighbours' animals, the graceful lyre form is common, and the horns are white to their tips, the most difficult of all the characteristics of Enyemibwa to breed. The Bahima herds show the result of determined and dedicated selection for beauty which is simply not evident amongst the cattle of the neighbouring tribes.[142]

This is not to say that Bahima had no interest in the amount of milk their cows produced, how healthy they were, or how they dealt with poor grazing or drought. The Bahima might have sacrificed production by refusing to give up their magnificent beasts for the zebu breeds, but they were still dependent on them for their livelihood. Occasionally, as I walked the hills of Ranch Ten or watched the herds grazing in Mburo, I would see animals that were very far from the ideal form, carrying small dark horns, a shrunken frame and patchy hides.

'Oh yes,' I would be told when I asked about this, 'I got hold of a zebu cow from Karimoja (or Baganda or Kenya), to breed to my Ankole bull, and this is her calf. I hope to get a good milker as a result. When I breed this one, her calves will look nicer but still give more milk, and within a generation or two the calves will look as beautiful as any other, but hopefully give lots of milk.' No one would know that their great-grandmother was an ugly cow from the east. Of course, the male calves could never be bred from and had to be got rid of. Only bulls known to be pure Ankole would give beautiful calves. The results of my genetic analysis showed that the Bahima had managed this trick of cross-breeding almost perfectly for generations.

For this analysis I collected hairs plucked from the ears of selected cows and bulls. Actually, however hard I tried I couldn't get close enough to do this myself, the horns sweeping past my head as the animals turned away from me. Fortunately, the owners had

---

142 The exception to this might be the royal family of Rwanda, who kept the famous Inyambo cattle, considered to be the most beautiful of all the breeds of long-horned cattle in the region.

Left: Other peoples also keep Ankole cattle, but do not breed them for beauty as the Bahima do

Right: Carefully managed breeding practices over centuries has created the Enyemibwa of Ankole

Left: A herd at rest during the heat of the day

Right: Ankole cattle must have large white horns to be thought beautiful

no difficulty, their beasts knowing them so well that they could stroke them on the neck and muzzle and tickle their ears before deftly plucking the few hairs I needed.

Once the DNA in the hair follicles had been analysed, I found that the genotype – the genetic material of the full complement of 60 chromosomes the animals carry – was a surprising mixture of both taurine and zebu genes. Compared against those of other breeds, it was no surprise to find that the closest relatives of my Ankole herds were other long-horned breeds, including the Watusi from Rwanda. What was surprising, though, was that the next closest match was a breed from the Lake Victoria basin in Kenya, the Kavirondo. Though a zebu breed, small and humped, it was nonetheless unmistakably closely related in genetic terms. This was unexpected, but it confirmed the explanation for the presence of the obvious crossbreeds I occasionally came across.

When the Ankole herds had been decimated by rinderpest in the late 19th century, the need to restock was urgent. The king attempted to raid cattle from nearby Rwanda but failed – the people there too had lost most of their animals to the disease and fought hard to keep those that remained. The genetic analysis suggests that he then turned to western Kenya, by this time under British control, to rebuild his herds. If this is what happened, the DNA shows that only cows were imported, not bulls. Despite what must have been enormous pressure to restock Nkore rapidly, my research showed that the easy remedy of importing bulls or breeding from the males of those first crosses with Kavirondo cows, had been strictly resisted.

Though the cows that I tested were thoroughly mixed up genetically, pointing to a history of mixing breeds, my tests of the Y chromosome of the bulls showed that these were almost pure taurine DNA. This confirmed what I had been told by my friends on the ranches; they would never breed from foreign bulls, because if they did they would never be able to breed back to cows that were beautiful. As long as they kept their bulls pure, whatever mixing had happened in the past or might happen in the future, the cows could always be bred back to look just as they should – beautiful long-horned ones. This was the Bahima's secret, and they had been at it for hundreds of years, perhaps thousands. Even when times were at their hardest, and the people were suffering, the principle of breeding only from pure Ankole bulls ensured that within a few short generations their cattle would be beautiful again.

The history of conflict between the Bahima and the authorities – first the imperial British, then the independent Ugandan governments, and finally and more narrowly, the conservation managers of Mburo – was understood, or at least described, by those authorities as being driven by competition over land and resources. They always put access to water and grazing at the centre of the conflict.

Looking at the cultural nature of the Bahima as a people, however, paints a rather different picture. Changes to the pastoral landscape of Ankole – set in train by the British, but driven onward by steady, incremental changes in the lives and livelihoods of the population – represented existential threats to the identity, way of life, sense of worth and values of the Bahima, and ultimately to their precious long-horned cattle. All these things had to be fought for, defended.

From as early as the 1930s, western cultivation began to convert parts of the landscape and resources that had been essential grazing for Bahima livestock, especially in hard times. The loss to them of the wetland areas and moist valley heads, where fresh grazing persisted longest, was especially problematic for them in practical terms, reducing the resilience of the extensive pastoral system. However, it was the curtailment of the free movement of their cattle by the spread of crops and fences that was especially hard for the Bahima to accept. For generations the Beautiful Land had been effectively reserved for the Beautiful Beasts. The ability to move fast, far and free was seen to distinguish Ankole cattle from others, and the same freedom of movement was part of being a muhima. Anything

Left: Cattle watering on the Rwizi River as it flows through the park
Right: A herd of cattle climbing the mists on a cold dry season morning

that interfered with this ideal of pastoral perfection, adherence to which acted to keep farms and farmers at bay, was to be opposed.

As the Bahima increasingly failed to counter these threats in the wider landscape, from the 1960s the battle came to be focused on Mburo. Mburo was in any case the land most valued by the pastoralists, and most associated with the values of their perfect pastoralism. Mburo was where the king had grazed selected herds, herds of pure bihogo, herds with horns like spears. These herds were the wealth of the nation in material terms and symbols of national prestige.

More than this, though, was that for the Bahima all the lands of Ankole were defined by the cattle that grazed them. The cattle and the land were inseparable. As mentioned earlier, before the name Nkore was taken, the land had been called Kaaro Karungi – the Beautiful Land.[143] Every hill and valley carried a name, many associated with historical events, others linked to myths that told the stories of the land and the people and the acts of the gods and the ancestors. The creation of Lake Mburo, for example, describes the sad tale of two brothers, one lost in the rising waters, one saving himself by escaping to the hills above.[144]

---

143 The Kingdom of Nkore, previously called Kaaro Karungi, took its name from events surrounding the defeat and death in battle of Cwamali of Bunyoro at the hands of Ntare I in around 1700. On learning of his death, Cwamali's mother is reported to have cried, '*Ebi shi ente za Kaaro zankore omunda*' (The cattle of Kaaro have killed the fruit of my uterus) from which the name 'Nkore' was drawn. Kaaro Karungi might have been a term of affection rather than a formal name for the nation. Kaaro is translated from the Ekyaro as 'land', but more formally should be understood as 'village'. Karungi is translated as 'beautiful' but probably meant 'good for cattle'. Today Kaaro Karungi is often translated as 'land of milk and honey', which gives a better sense of its significance than 'village good for cattle' would.

144 Two brothers, Mburo and Kigarama, occupied a wide valley lying beneath a steep range of hills. Kigarama was warned in a dream that a great flood was coming and that he should leave the valley with his family and herds and climb the hills above. Mburo, however, refused to listen and perished in the valley. Today the lake is named for him, and the hills are named after Kigarama.

Land with water and grazing but without cattle was an abomination to the Bahima; in the Bahima worldview such land was without meaning. Mburo when declared a conservation area and kept free of cattle, could not stand – could not be *allowed* to stand. It was to no avail to talk to the Bahima about the herds of impala and zebra and the importance of protecting them. The Bahima had no particular dislike of these animals that shared the range with them, but they had no particular interest in them either. Even if they had, the lack of cattle on the land would have remained an insuperable problem.

Unfortunately for the Bahima, those who controlled Mburo held diametrically opposed views. For them, Mburo was for wildlife, full stop.

It becomes clear that although the concept of Mburo might have represented a foreign ideal drawn from the west, it was an ideal that had been internalized, one that defined the identity and worth of Uganda's conservationists, and which was pursued with almost fanatical energy by park managers. The sight of a cow was like a personal affront to them. The smell of the dung and the trails the cattle trampled through the bush were loathed with a deep and bitter loathing.

When cattle were barred from the game reserve in the 1970s and even more so, during the early 1980s as the national park was established, the conservation authorities went after the cattle with a vengeance, chasing them from the land, impounding them and even shooting or spearing them. The Bahima responded by grazing their animals at night, fleeing from patrols and, when there was no other recourse, bribing the rangers. This uncomfortable situation continued for decades, carried on despite the efforts of our Community Conservation project, and no doubt continues today. The Bahima's fight to ensure the presence of their cattle on the land that was Mburo and the authorities' insistence on their absence from it represented entirely different sympathies with the land.

So beneath it all, it was the different meanings attributed to the terrain defined as Mburo that defined the terms of the battle.

Our failure to engage the Bahima had served to demonstrate to me that regardless of our good intentions a different approach was needed. If what underlay the conflict between the Bahima and the park was the incompatibility of the two different worldviews, then that was what we needed to address. If the conflict was to be resolved something had to give. If the problem was fundamentally the product of a clash between cultures and values, perhaps it was time to start looking at solutions of a cultural nature.

6

The nature of conservation

When I started out in nature conservation, I saw it as entirely positive and unquestionable good. There could be no considered resistance to it when, both at home and abroad, the beauty and wonder of the natural world was vanishing before my eyes. The righteousness of the endeavour seemed incontrovertible. No questions, caveats or controversies hung over it, and when I first set out for Africa no moral difficulties intruded. Forty years later, after work in Africa, southeast Asia – and now, in the landscape of my childhood, Ashdown Forest – the day-to-day reality of delivering conservation forces me to recognize its complexities and moral ambiguities.

On arriving in Africa in 1981 I had learned that the continent was home not just to wildlife but to many people. Forty years later there is less wildlife but many more people.[145] A multitude of questions require me to consider conservation in a more nuanced light than when I first set out to save the world.

It didn't take long before I had to acknowledge that the fundamental concept of the national park, which lay so squarely at the centre of my ambitions, was flawed. Conservation might be a good thing – indeed, I believed strongly that it was – but as practised through the creation of strictly protected areas, it was certainly not good for everyone. There was no escaping the reality of the sometimes terrible costs associated with parks and the wildlife they protected. The endeavour I had signed up to was just not fair.

I was not alone in recognizing this. By the 1980s conservationists had already understood that setting aside areas for nature imposed costs on the poor and powerless. As well as losing access to their land and resources, people suffered from real depredations of wild animals. Unless we took this seriously it seemed inevitable there would be trouble

---

145 Uganda, which lies at the centre of this story, has one of the highest rates of population growth in the world. During the 35 years that have passed since I first went there, the population has grown by over three times, from a round 13 million in 1981 to over 41 million today. For most of the time I have worked in conservation, it has been almost taboo to talk about population growth as an issue. The significance of this dramatic and rapid increase in human numbers for efforts to conserve nature in Uganda cannot be denied, though, and there are now several programmes linking conservation projects to working with communities on social and reproductive health.

in the future. If the actual costs and the potential benefits from conservation were not distributed more fairly, it seemed unlikely that those meeting the costs without getting the benefits would continue to tolerate our parks or support our idea of conservation. It seemed more likely that at some point they would rise up against their governments, against conservation organizations, and against people like me, and the parks would be swept away. The events at Mburo – events that fomented a mini-disaster for communities and conservation alike – attest to my misgivings.

The analysis of the costs and benefits of conservation, a somewhat detached way of looking at the reality of protecting nature, was just one part of the picture. That some people were losing out as a result of a notion I was working to bring about was one thing, but to acknowledge that I was gaining from it personally was another. I became aware that it could be, and indeed was, argued that I reaped the benefits of a job, a salary, and a four-wheel drive to cruise the bush in. This was all true – but I was aware, too, that these were not what had induced me to travel to Africa. When I considered my good fortune to be living in such glorious landscapes, so full of the sights and sounds and smells of wildlife, the material remunerations of my position were not the values that sprang to mind.

What excited me was being in the wild haunts of the wild creatures. It was the pleasure I got from climbing to the top of the Kigarama Hills, scrambling across dusty shale; from scanning the papyrus fringes of Lake Mburo below for buffaloes in their wallows; from sitting in the indigo shade cast by the velvety leaves of a Combretum tree. It was the thrill of startling a solitary eland wending its slow way along a ridge-top through grasses heavy with seed and dotted with the compact crimson or delicate yellow of hibiscus flowers. It was the bottled excitement of creeping along the edge of a swamp to watch a saddle-billed stork pointing its improbable red, yellow and black bill like a poised spear.

Left: Lake Mburo on a hazy morning

Right: A fisherman mends his nets on Lake Mburo (copyright Connie Bransilver)

136

Left: Hibiscus cannabinensis
flowersand carefully peeled twigs
Right: Sleek young male impala

Goliath heron scans the lake edge for prey

It was the quietness that settled over me as I dipped my oar into the still water of Lake Mburo at dawn, slipping through the thin mist close to the shore looking for crocodiles. A restrained splash and widening ripples would tell me I have been spotted and the croc is gone. The comical, ungainly scamper of a finfoot fleeing across the water on flapping feet is adequate compensation.

These were the rewards of my job; these were the benefits I sought and received. Such experiences which nature gave to me explained my determination to protect it, and justified, in my mind, my presence and actions. These were the recompense for my efforts.

I knew that these perceptions were particular. They might even have been unique to me. It was hard, then, to suggest that such experiences might be relevant to the sharing of costs and benefits of conservation. Nonetheless, when I considered what I got from nature, I began to wonder if our neighbours, those people we wanted as allies of the parks, might also consider their own experiences of nature as benefits too. The experiences of and connections to nature of the people living around Mburo were likely to be different to my own, but it began to seem sensible to consider how we could make them part of our efforts to describe the values of the park.

To go down that path, though, conservation practice would have to consider the values of nature much more broadly than it did, and work to actively include those of our neighbours in its prescriptions. If we conservationists could find a way to open up our thinking, though, we would be able to include all sorts of benefits and values in our efforts to define and describe our work and use to reach out to others. Rather than focusing narrowly on balancing economic loss with economic gain, we could also think about the intangible values our parks protected, and provide access to a wealth of benefits that, though not material in nature, might be no less important to our neighbours.

As my thinking about parks and their management matured through my experiences at Mburo and later in Vietnam I began to review their positive results for nature against their negative impacts on local communities. When trying to understand why protected areas succeed or fail, the rise, fall and revival of Mburo seems like a good example to consider. I have discussed the area's special connections to the Bahima people, its history of protection under different guises, and our efforts to recover it from loss. How did this history of conflict and reconciliation affect Mburo and values it was established to protect, and the communities living in and around it? What did the actions taken in the name of conservation achieve, and what can the outcomes tell us about protecting nature more generally?

Two perspectives exerted a strong influence over the community approach we had developed at Mburo, and both, in different ways, affected how we embraced the challenge of rescuing the park. It was the notion that wildlife must 'pay its way', that the financial value of nature provides conservation's most immediate and primary justification, that had the most direct impact on our programme. Equally important but less concrete in its influence was the idea that scientific explanations of the natural world and its workings provided justifications for conserving it. The concepts of biodiversity and ecosystems increasingly dominated consideration of what was important about nature and, therefore, why it had to be protected.

This thinking, I believe, represented a shift in the role of science in conservation. When I was a student in the 1970s, science seemed to be understood as a powerful tool used to design and deliver conservation programmes. But as my career progressed, science began to determine why we undertook them. Rather than informing us *how* to protect nature, science began instructing us *why* we did it. Together, economics and science insisted that conservation was neither about what people might consider desirable for an abundance of reasons, nor a choice that could be agreed though an exchange of ideas, but rather an essential for human survival and therefore not a choice at all. Further, science and economics were presented as the only tools that could deliver conservation successfully.

Throughout the 1980s and 90s these injunctions were presented as unchallengeable arguments for conserving nature. In parallel, however, protected areas like Mburo were increasingly castigated by advocates of human rights. Protected areas had been taken from people and the land should be returned to them. Voices raised in support of Indigenous Peoples pointed to wrongs done by protectionist conservation around the world, including in Uganda. Parks were described as foreign constructs that robbed people of their rights to hunt and gather, to graze and farm. Parks robbed people of their rights to pursue traditional ways or life, hold their beliefs and enjoy their culture.

In 2008 the term 'green grabbing' was coined: the expropriation of land by governments, powerful international organizations and even private individuals who bought sometimes vast tracts of land to manage for conservation. Legal challenges were mounted on behalf of communities dispossessed, forcing some governments to make concessions.[146] Rights-based approaches argued that there was no need for governments to be involved in conservation at all, as the real owners of the land would protect it; numerous sites around the world were conserved by Indigenous Peoples and local communities, and had been for centuries.

Decades before this, Arthur, Moses and I had recognised the problematic nature of Mburo as an institution, and we were acutely aware of the problems it caused our

---

146 Legal cases brought against governments in the United States of America, Canada, Australia, New Zealand and South Africa, and more recently in Colombia and other developing nations, have succeeded in forcing accommodations with Indigenous Peoples that recognize their rights while continuing to manage lands for conservation through various agreements. The Batwa in Uganda are currently being assisted to sue the government for the return of their rights to the forests of Bwindi, Semliki and Mgahinga, through at the time of writing I know of no progress made in resolving the issue.

neighbours and hoped-for partners. And we were shamed by the history of its gazettement. But we did not question that the land should be for protecting nature, continuing what had begun 60 years before. We justified this in part by reference to the context and times we were working in. In the 1990s, Uganda National Parks, the donors funding us, and the African Wildlife Foundation I worked for would have refused to discuss the legitimacy of Mburo itself or of national parks in general. They were committed to protect the park. We were, too – but we believed a more acceptable way of doing it was needed. I was sure that Mburo's many values would not survive without government protection and our project, but I wanted to open a discussion about what protecting Mburo really meant locally.

This is not to say that there had not been voices raised against protected areas in general and Mburo in particular, even after two thirds of it had been cut off. Throughout the decade when I focused on strengthening Mburo, I was on the receiving end of challenges to its validity and accusations of perpetuating a damaging colonial construct. I would point out that Uganda had international obligations to protect its heritage, and that whatever their origins, national parks were embodied in Ugandan law.

Whatever the arguments, the Community Conservation project required us to recover the park as a place of conservation. It was implicit that the status of national park was to be retained on the part of it that remained. At the time the powerful local players arguing for its dissolution were not kind-hearted folk determined to bestow land on the community. Nor were they anxious and determined to conserve the land, the wildlife or other values for their nation. They were very obviously pursuing individual gain, focused on personal, not communal, ownership of the land. I felt ready to oppose them, and indeed compelled to do so. To Arthur, Moses and me there was no question that the very real injustices perpetrated when the park had been created had to be addressed – but we nevertheless believed we could do that while protecting Mburo.

Despite my growing misgivings about the nature of parks and the challenges they posed, at the time when the Community Conservation project began I was convinced of the need to protect nature for the contributions it made to the quality of lives, mine included. By this time, I had worked in several protected areas in Africa, and had visited more. I had seen that they were richer in wildlife and other natural values than the lands surrounding them. Unprotected lands were affected, inevitably, by high levels of conversion and economic production. At Mburo throughout the 1990s I watched the rapid changes to the land that had been cut out of the park as it was settled, fenced, cleared and farmed. We had little doubt that without protection this was what would happen to the rest of Mburo.

Unless land is set aside from 'business-as-usual' demands to exploit it, nature will be the loser, and setting aside places to protect nature means that rules for managing them are essential; defining acceptable levels of change inevitably requires decisions on what can and cannot not be done within them, and to what extent. Without this it seems evident that the beauty and diversity of the natural world will be steadily lost. Even with the global network of protected areas the losses continue, and without these areas I am convinced that our ability to slow these losses will vanish.

'Business as usual' in many developed and developing country communities and corporations will put land and resources to their most profitable use, normally in the short term, over four or five years, as when a forest is chopped down for timber. But it can also happen incrementally, as when farms expand across a landscape over generations. In any case every natural area will be turned to some other and more financially profitable use. At Mburo this would be banana plantations and intensive dairy production. On Uganda's Sese Islands, it might be palm oil plantations, just as it was on the islands of Indonesia. Paradise will be paved, to 'put up a parking lot, with a pink hotel, a boutique and a swinging hot spot' or more likely, shopping malls and industrial parks.[147]

Apart from the protected areas that allow and even encourage uses that are compatible with the protection of nature, my experience suggests that in the absence of limitations land use will generally result in the maximization of production and profits. Unless nature provides the most economic use, as it does in a few places, little space is left for nature. This is not to say that nature cannot and does not exist in lands managed for production; just that as production increases so does nature inevitably seem to decrease.

Whether in Scotland's or Oregon's temperate highlands, or the tropical lowlands of Borneo or Cameroon, natural forest *can* be managed for timber production in such a way as to retain its birds and insects and other plants. Production forests *can* retain levels of

Left: Palm oil plantations in Indonesia often replace biodiversity-rich forests
Right: Rubber plantations in Sumatra, Indonesia, with trees planted in regulated rows

147 Was more truth ever put more sweetly, sadly and succinctly than by Joni Mitchell in her 1970 song 'Big Yellow Taxi'? It goes straight to the point – 'They paved paradise, and put up a parking lot'.

complexity and richness similar to those protected from logging. Plantations, however, whether pine, rubber or oil palm, are by comparison sterile, silent places, their regularly spaced trees weeded to ensure the best conditions for production as they grow steadily towards the day when they will be felled, leaving a bare scar awaiting the next crop.

Some studies suggest that over time diverse forests can be more productive than monoculture plantations.[148] In general, though, area for area, year for year, plantations seem to outperform forests in terms of cubic metres of timber or litres of palm oil, and certainly in terms of profits. Corporations, governments and even small-scale farmers continue to invest in clearing natural forests for plantations, voting with their machetes as it were. Most private landowners, working to feed their families, build financial empires and pay their shareholders seem prepared to live without the rumble and munch of elephants, the melodies of birds and the buzz and flutter of insects, so long as they can hear the rustle of money.

Protected areas keep a place for the elephants and insects and all the other species of diverse ecosystems because these are what they were established to 'produce', or, rather, to protect. National parks were not established to meet economic targets or compete with timber or palm oil to justify their existence, at least not when first conceived. America's National Park Service was founded for a very different purpose from that of the Forest Service, though both were established at the same time, in the early 1900s.

Things change, though. It was not long after I started work at Mburo in 1991 that I sensed changes in the language used to talk about conservation; commercial and business terms began to proliferate in our reports and proposals and the way we advertised our organizations. By stealth, it seemed to me, economics began to dominate our analysis of nature's values. Increasingly we couched our arguments for conserving nature in economic terms and communicated its values in financial terms.

So throughout the 1990s our project undertook all sorts of activities to identify and describe the economic values of Mburo, then use these to try and build bridges to our neighbours. We sought to win their support through creating and sharing tangible benefits. It would have been hard to contest the immediate rationale for this approach; our interventions were designed in the spirit of good neighbourliness, of sharing, and of working together to reduce the difficulties of communities. Contributions to the wellbeing of our neighbours was tied to the future of the park. It had taken us some time to recognize that Mburo's presence contributed to local poverty, but once we had accepted that, it became a priority for us to reduce this and other negative effects.

---

148 The time frames involved and the complexity of the species in and physical structure of different forest types make it hard to study productivity under different management regimes. However, a global assessment of the relationship between the number of trees species and the production of timber found that more diversity was linked to higher production. Whether this relationship would be enough to convince foresters that the gains from diversity outweighed the costs of making operations more difficult is uncertain; many factors affect the profitability of production and the decisions taken by commercial foresters.
'Positive biodiversity-productivity relationships predominate in global forests', Liang et al (2016) *Science*, Vol. 354, Issue 6309, pp. 196–207.

The desire to rebalance the costs and benefits of conservation was certainly right – but unhappily, efforts to achieve this became synonymous with the simplistic notion that it was through poverty that communities were driven to poach, steal timber and graze animals in parks. Accepting that protected areas should be part of efforts to relieve local poverty, and that conservation had a special responsibility to minimize the suffering it caused, was right in its instincts and intentions. But linking nature conservation to economic development so directly would, it turned out, take conservation down a path that had unforeseen – and disturbing – consequences.

As Moses went about implementing activities to increase households' incomes and reduce their dependence on natural resources, two outcomes emerged which I regarded with qualms. Firstly, our replacement of the resources that people had looked to Mburo to provide diminished its role as a provider of things people wanted, so the park became less relevant to their interests. This was even more clearly the case in other parks, particularly the newly established forest parks. At Bwindi National Park, steps were taken in the 1990s to provide local communities with key resources, especially firewood and building poles. For centuries people had turned to the forest for those and other needs, so to halt the continued dependence on the forest, communities were assisted to grow trees on their farms. They were also encouraged, for a limited period, to collect seeds and cuttings of plant remedies to grow in their gardens. The intention was to reduce the temptation for people to go into the forest – but to me, even back then, making the forest less relevant, which these substitution projects inescapably did, seemed unwise.

Rather than substitution, I liked the idea of activities that would deliver people's wants while preserving or even creating new connections to the forest. Bee-keeping is a good example. The quality and quantity of honey from hives located outside the forest depend on bees foraging for nectar and pollen inside it. 'Butterfly farming' also links people to forests in a productive way. The butterfly houses, vast glass palaces that deliver the experience of tropical butterflies to human visitors, depend on regular supplies of butterfly pupae. Communities supply these by growing food plants that lure butterflies out of nearby forests to lay their eggs. Caterpillars hatch, feed, pupate, and are collected and sold to the butterfly houses. To me this seems a perfect model of a healthy and sustainable dependency on forest; the forests support the butterflies that provide the pupae that generate the income.

My second concern with the financialization of conservation was that it obliged parks to be described in economic terms, thus encouraging, even demanding, that our neighbours and their leaders see the parks in that way too. Parks were presented as engines of local development. At Mburo we would describe the money that would flow into the local economy and the number of jobs that would be created once tourists started visiting. We encouraged people to look east to Kenya or south to Tanzania to get an idea of how the park would change their lives.

Of course, thinking of national parks in relation to tourism wasn't new. Uganda's booming tourist industry in the 1950s and 1960s had not been forgotten, and some of the most exuberant arguments for making Mburo a national park had been its tourism potential. Even the first national parks established in America had been promoted as destinations for tourists, and some had even been established in partnership with the newly flourishing railways, ready to carry visitors from the cities to the wilds.

I understood very well why we promoted Mburo as a source of revenue for the nation and for our neighbours – but it worried me nonetheless. And as it turned out the problem was, as described earlier, that we just couldn't deliver.

So how else was Mburo to drive local development? The park did provide some resources, legally or illegally, especially fish, but ensuring use was sustainable meant that the harvests would actually have to be reduced, hardly a practical demonstration of the park's value to local people. To counter this we resorted to suggesting all sorts of financial values to the multitude of species in the park, implying that wealth would flow from them at some point.

We talked so often of the value of, for example, the Madagascar periwinkle, a small plant lauded for its cancer-treating properties, that if asked to describe why Mburo was important, people would bring up the subject of this plant, though it was unknown to them and didn't even grow there. The Madagascar periwinkle might have created fabulous profits for pharmaceutical companies, but it did nothing for the people of Madagascar, or for the conservation of the forests in which it is an endemic species. It was certainly not going to help the communities around Mburo.[149]

Experiences like this, where over-optimistic rhetoric was fed back to us, fuelled my worries about extolling the financial rewards of protecting nature. It seemed we had hung a dangerously double-edged sword above our protected areas. A colleague, describing an experience in Zimbabwe, perfectly captured my concerns.

One of Africa's most highly regarded conservation and development projects went by the catchy acronym of CAMPFIRE, a wonderfully resonant title standing for the slightly more clunky Communal Areas Management Programme for Indigenous Resources. CAMPFIRE had great success in encouraging communities to make use of the rich wildlife on their communal lands rather than converting it to farms or ranches. It was argued that building tourist lodges, selling hunting licences and culling wildlife would give the best returns from their otherwise marginal rangelands. CAMPFIRE

---

149 The therapeutic properties of the Madagascar periwinkle (*Catharanthus roseus*) were well known to traditional healers, and it was this knowledge that attracted interest in it from the pharmaceutical industry. Two important cancer treatments, vinblastine and vincristine, were isolated from the plant. Despite their importance and the profits made from them, little if any benefits reached the people of Madagascar, where the species is a native endemic, or its forest habitat. Not only that, but although the Madagascar periwinkle is an invasive species found all around the world, it is now considered endangered in the wild. https://www.cabi.org/isc/datasheet/16884

achieved great success, and many communities garnered substantial financial rewards from their wildlife.

All was well until a new player appeared on the scene, selling a package of finance and support for planting high-yielding cotton. Cotton would give a better financial return at guaranteed prices than the uncertain wildlife industry. My friend found himself in meetings trying to persuade communities not to tear down their wildlife businesses to plant cotton. Not only were they getting healthy economic returns from their wildlife, he explained, but they were part of a long-term and sustainable use of the land that protected the bush in all its beauty and diversity, protecting the national heritage and ensuring that their children and grandchildren would be able to enjoy a life with wildlife.

He had little response, though, to the villagers' quizzically raised eyebrows as they repeated, as if to a child, that planting cotton would give them better financial returns. CAMPFIRE had urged them to keep their wildlife because it was their best financial option. But now they had a better option. In terms of the arguments they had accepted earlier, it made complete sense to replace wildlife with cotton. The blade of economic argument had slipped and cut the hand that wielded it.

While we were struggling to start canoe safaris and set up nature trails, gorilla trekking, officially launched in 1993, was just beginning at Bwindi. Copying Rwanda, where gorilla tourism had become a significant business, Uganda embarked on a process to habituate troops of mountain gorillas. The high demand for the trekking permits led to a scramble to build lodges for the tourists beginning to arrive. Tourism revenues were soon of sufficient importance that the fledgling industry was strong enough to fight off proposals for a road to be built that, running through the heart of the forest, would have cut the gorilla habitat in two.

Gorilla tourism has grown steadily since those first visits. At a charge of US$700 per visit, permit sales were estimated to be worth US$25 million in 2018/19. Assessed as a whole, tourism is Uganda's second largest earner of foreign exchange and employs over 200,000 people.

Despite this, the benefits reaching local households are limited. Certainly, some communities in the locations where tourists stay in or pass through manage to trap some financial rewards, but most of the millions of

Mountain gorilla tourism is important in Uganda and Rwanda

people that live around Uganda's parks, including those around Bwindi and Mburo, do not. Not only do most of the profits go to the operators, often foreign, but also, however big the industry gets, the size of the populations around the parks causes the benefits that are shared locally to pale into insignificance as far as any individual is concerned.

Conservation cannot ignore economic realities, whether these seem to support or undermine its objectives. Parks are expensive institutions. Building and maintaining roads, gateposts and ranger stations is costly, and rangers and wardens are needed to patrol, monitor, engage with local communities and much else besides. The costs of the parks in terms of lost opportunities – the gap between what the parks produce compared to what the land might produce if, for example, put to cotton, cattle, palm oil, or even parking lots – are higher still.

Though not always the case, the continuum from full production at one end to full protection at the other influences the profits that can be extracted from a piece of land. A Category 1a nature reserve, from which even tourism is banned, earns very little, and though efforts are made to describe more fully the value of what such reserves contribute to humanity, very little of this can be turned into money and jobs for local people.

And rightly or wrongly, it is money in people's pockets that we have promised. Parks are expensive in opportunity costs terms, and once again it is local communities who carry most of these, as it is largely their opportunities that are lost or constrained. Add the costs of wild animals damaging their crops and the costs of protecting them – children spending days in the fields rather than in school, for example – and it is easy to see why so many people living around protected areas are less than enthusiastic about them.

The costs that conservation levies on communities, often poor already, cannot be ignored. There is a moral demand to engage with this fact, but there are practical reasons too. Crops of maize, beans and bananas grown around Mburo are damaged, sometimes extensively, by buffaloes and bushpigs coming out of the park at night, while monkeys and birds descend on them during the day. Park staff try to respond to these depredations but are far from successful, and the anger of farmers waking in the morning to discover devastated fields is a heavy counterbalance to hard-won positive relations achieved by community programmes. The losses suffered by farmers around Mburo echo the losses experienced by communities living around protected areas the world over. So too do their feelings of frustration and anger – inevitably, and not unreasonably, aimed at their local protected area.

Nor should we ignore the fact that wildlife can be dangerous. At Mburo the main threat to human life and limb is from buffaloes and hippos. In other places it might be elephants or tigers. Loss of life might be rare, but attacks do happen, and unsurprisingly they have a powerful effect on how people regard efforts to protect potentially dangerous animals.

In South Africa, I interviewed 300 people living on the edge of the Umfolozi and Hluhluwe nature reserves. The reserves contained lions, and despite the fences almost everyone I spoke to lived in fear that the lions would escape. No one had actually been attacked, no one knew anyone who had been attacked, and no one said they even knew of a specific attack. But all my respondents believed that attacks happened and would happen again, perhaps to them or their family. Though the managers of the reserves made light of these fears, branding them irrational and unrelated to the real risk, it was the perceived risks that nonetheless influenced how those communities viewed their local protected area; and their view of it was not positive.

In 1988 I spent a year working in southwest Cameroon, in the Korup Forest. A national park was being developed and I was there to assess how this might affect the many villagers living in and around the forest. Not surprisingly they had plenty of concerns, but one of the things that worried them most was the protection of elephants. It was not that they hated or feared elephants *per se*, but they disliked the uncertainty they caused. A simple visit to relatives in the next village or a trip to harvest cocoa from their fields was plagued by the possibility of encountering these behemoths. In their view, elephants, like some misplaced dinosaur from the past, represented the very opposite of what they wanted and expected from the development of their country: order and predictability in their lives.

Back home in England, my friends would lament the fate of Africa's elephants, slaughtered, they believed, by cruel and cynical Africans. While far from defending the killers of elephants or attempting to present the complexity of efforts to protect them, I would nonetheless suggest that there was room to sympathize with people who might rather not have elephants around.

'Imagine,' I proposed as we our sipped pints in the pub on a fine summer evening, 'Imagine taking a short cut home through the woods and across the fields. It is getting dark now, and you know an elephant, a towering, grumpy lone male perhaps, might loom suddenly from the gloaming. It charges with a terrible crashing of saplings and fence posts as you try to flee, your boots made heavier by beer-filled legs. Or it might simply fade silently into the soft night on its great cushioned feet. Either way, you might feel rather differently towards elephants.'

Implementing the community conservation project's two phases from 1991 to 1998 I found that our talk of Mburo's importance sounded increasingly hollow. Again and again I found myself explaining that it was in the local people's best interest to support the park because it supported them. I talked about how the park supported local interests; the community did not need to love the park, they just needed to see what side their bread was buttered on. But I knew, too, that the interests that had caused Mburo to be turned into a national park had been all about protecting nature. So there was a gap between my rhetoric and the reality that the communities lived with.

Our project's best efforts to build economic bridges to our neighbours came, of course, to little. This was partly because the apparent success of some initiatives was not in truth based on the financial values of the park, but on monies channelled through the project. In the end, we were forced to explain that though the benefits received were not really produced by the park, they came nonetheless because of government and global interest in Mburo and its wildlife. This was not what we had intended when we embarked on our efforts to demonstrate how the park contributed to local wellbeing, but it was at least more honest.

Whatever the truth underlying the economics of Mburo and other parks, the economic arguments for conserving nature are now so prevalent and so accepted that they seem to obscure other perspectives. The conservation movement has actively engaged in redefining nature either as a commodity, like oil or grain, to be bought, sold and speculated in like any other, or as a service that can be supplied and charged for. Where this financialization of nature will end is unclear, but it seems unlikely it will help deliver benefits into the hands of poor communities living around the protected areas or help deliver on the promises that the parks would benefit their neighbours. The opposite, in fact, seems more probable, because the economic value of nature is drawn off through returns on financial investments made by outsiders.

Jean-Jacques Rousseau, writing in the 18th century at a time when the enclosure and privatization of land was proceeding apace in France – land grabbing as we call it today – ironically noted:

> The first man who, having enclosed a piece of ground, bethought himself of saying *This is mine*, and found people simple enough to believe him, was the real founder of civil society. … From how many crimes, wars, and murders, from how many horrors and misfortunes might not any one have saved mankind, by pulling up the stakes, or filling up the ditch, and crying to his fellows: "Beware of listening to this impostor; you are undone if you once forget that the fruits of the earth belong to us all, and the earth itself to nobody."[150]

His warning against privatizing the fruits of the earth – the benefits nature provides for us – can still be heeded. Enormous inequalities have resulted from the privatization of land, and the process of land grabbing continues, in Uganda as well. The consequences of privatizing the benefits of nature are likely to be even more severe. Tied to an out-of-control global financial system, it seems certain to result in very undesirable outcomes for most people, if not outright catastrophe.

---

150 Quoted from Rousseau's *Discourse on the Origin and Basis of Inequality Among Men*, published in 1755.

If economics has taken a worrying direction, science too seems to have veered off the tracks as far as conservation is concerned. While developments in health, agriculture, energy and information technology, to name a few, bring enormous benefits to humankind, these same advances increase demands for land and resources, pollute the air and water, change the global climate and threaten nature and its systems. Scientific advances are also increasing our understanding of how nature functions, and it is increasingly central to the design and delivery of conservation initiatives. However, these powerful capacities, by describing what nature is, are beginning to dictate why it must be protected.

Ecology is the study of the interactions between living organisms and the environment in which they live. Like other species, humans respond to the animals, plants, weather and peculiarities of location that make up their environment, and have done so since before we became humans. Today, hunters, gatherers, pastoralists, farmers, people that work the land and its resources, not to mention naturalists and nature lovers of all kinds, have built a profound store of ecological knowledge.

Some of this wisdom is in the form of myths and stories, sayings and old wives' tales. Some can be found in publications that, old and new, contain ancient and modern knowledge. Almanacs to guide farmers have been published annually in Britain from the mid-1600s, and *The Old Farmer's Almanac* has been published in America since 1792, making long-range weather forecasts using secret systems comprising observations of planets, animals, plants, clouds and much else.

Employing science to understand how nature works has been part of modern conservation since 1929, when George Melendez Wright, a young naturalist working in Yosemite Valley National Park, carried out the first systematic surveys of wildlife to guide park management. The design and management of early protected areas focused on landscapes, iconic species, and the practicalities of establishment and policing. Natural features, rivers, lakes, shores and mountain ranges were used to define boundaries, and by a combination of luck and good sense these often made ecological sense too.[151]

As conservation increasingly focused on biological diversity, and on rare and threatened species in particular, ecology became more important in the design and management of protected areas. With the growth of ecological knowledge, it became evident that there was more to conserving nature than simply eyeballing a forest or a range of hills and deciding to protect it. That animals needed large areas to survive in

---

151 Unlike many of Uganda's protected areas, which had been declared in the early 1930s. The Rwenzori Forest Reserve boundary, for example, took the simple expedient of following the 8000-foot contour. As the agricultural boundary was well below this altitude at that time, the new boundary line cut straight through the forest. In contrast, the Mabira Forest Reserve was established in an area that already had villages and cultivation. In this case the boundary was simply drawn around them; it makes no reference to the ecology of the forest and is extraordinarily difficult to maintain.

was not new or surprising knowledge, nor that some animals travelled long distances while others had fixed migrations. It was now understood, however, that parks should try to accommodate these requirements. As research into the behaviours of individual species and the ecologies of different ecosystems began to reveal their secrets, demands for science-based conservation grew.

Relationships and interdependencies between species are not easy to observe or analyse, and predicting how changes to their environments will affect them can be hard. Efforts to protect a species or ecosystem can easily founder without sufficient knowledge. The sciences of ecology and animal behaviour are thus important tools for fine-tuning the design and management of protected areas, for monitoring their integrity and checking whether something might be happening, and for helping to understand why and what might be done to remedy the situation.

The first time I was directly involved in applying science to conservation was as an assistant on a project studying suni antelope in South Africa in 1983. At less than 40 cm high, the suni is smaller than most domestic dogs and some cats. It inhabits the dry coastal bush and forest that runs in a narrow band from Kenya to South Africa. Shy, nocturnal and favouring thick bush, it is rarely seen. Nonetheless, some managers of reserves in South Africa began to feel there were fewer of them around. When surveys confirmed that their numbers were indeed low, a study was commissioned to find out why. I was fortunate to spend six months in False Bay Park,[152] a narrow band of forest growing on sand dunes along the coast, helping with this research.

I walked transects through the forest to count the leaves and fruits falling from trees that the suni would hoover up with their mobile mouths. I plucked every leaf from every plant growing in cubic metre plots and 'bombed' them – burning them to carbon flakes – to estimate the nutrition available to the suni. I carried car batteries on my back to power spotlights to follow the suni at night to see what were up to. I learned to identify trees, shrubs and grasses and discovered how difficult ecological processes are to understand. The suni, it turned out, had plenty to eat but were struggling to find the patches of really thick bush they needed to hide out in. This was apparently because the nyala, another browsing antelope, were steadily thinning out the forest floor, preventing thickets, the shelters the suni needed in order to survive, from forming.

Park managers must respond to all sorts of challenging interactions between species if conservation objectives are to be met. In the case of the suni and its disappearing refuges, the response seemed relatively clear, if not easy to implement. The nyala were

152 At the time called False Bay Nature Reserve, this protected area lies on the edge of Lake St Lucia, a large estuarine ecosystem on the coast of KwaZulu Natal.

Suni antelope, Neotragus moschatus, *in False Bay Reserve in South Africa*

culled. Unpalatable though this might be to some, it is not uncommon to artificially control animal numbers to try to promote the survival of one species or habitat against the impacts of other species, especially when there are no predators available to do the job. This is especially challenging where it involves large and charismatic species such as elephants, whose killings tend to stir powerful emotions in humans.[153]

Interactions between species at Mburo threw up some tricky issues. Learning my way around the park's natural history in the early 1990s, I found that we were lucky enough to have brown-chested lapwings. This small bird, with unobtrusive yellow wattles and chestnut markings on forehead and chest, is pretty rare in Uganda, and Mburo was the place to find it. For serious birders, and indeed for biodiversity-fixated conservationists, Mburo and the

---

153 The culling of elephants in southern Africa, undertaken to protect forests from these voracious feeders on trees, is highly controversial, and has been replaced by some conservation authorities with elephant contraception. Both culling and contraception are means of controlling elephant number in circumstances where it is felt control is needed. In Britain we have similar though perhaps less difficult concerns over the impact of native deer or domestic sheep preventing forest regeneration by eating every seedling that puts its delicate leaves above the soil.

brown-chested lapwing had become synonymous. When describing Mburo to investors, tourists and even local communities, I was in the habit of mentioning that Uganda's only impala were found there, that more mongoose species were found there than in any other park,[154] and that the presence of this otherwise rather unremarkable bird made Mburo special.

Over the course of several years, though not a serious birder I realised I was seeing the lapwings less and less frequently. In the end I had a reasonable expectation of finding them in just one spot. Under a handful of very large acacias, hippos had grazed the pastures down to a lawn as fine as a golf course green, scattered with patches of sandy soil. This, it turned out, was the perfect habitat for the lapwing, and it was disappearing elsewhere in the park because when the Bahima and their herds were removed there were no longer enough grazing animals to keep the grass low and thin enough for the lapwing. If I wanted to see them, I had to look outside the park, where the high numbers of cattle were more than able to keep on top of the growing grass.

This was a conundrum of the sort that park managers still have to deal with. Hanging onto the lapwing demanded that its favoured habitat be maintained, but removing the cattle was causing this particular habitat to disappear, and the bird along with it.[155] Given time, higher numbers of buffalo would probably solve the problem, but it might be too late for the lapwing. While science helps us understand such issues, and can provide ecological solutions, it cannot determine what actions we should take; our decisions must be based on what we are trying to achieve, in this case protecting a particular species. Science cannot help us with this.

And herein lies a problem. As the complexity of ecosystems is becoming understood better, a set of interlinked perspectives has come to be broadly accepted within the conservation movement. Perhaps the most influential is the idea that the complexity of natural systems makes them incredibly delicate. Because everything is connected, it is argued, every part is essential for the survival of the whole. Keystone species hold these structures of interconnected species together, just as a keystone holds up an arch. The removal of keystone species – perhaps even *any* species – could result in a cascade of lost connections leading to the collapse of the entire system. A tropical forest with its millions of trees and fungi and insects and birds might fall into ruin as a result of the loss of any one of them.

---

154 Six species, including the slender mongoose, not recorded in any other of Uganda's parks.

155 When England's large blue butterflies were surveyed and found to be disappearing, urgent efforts were made to protect the few sites in which they could still be found. Not understanding enough of the butterfly's lifecycle, the people who made these efforts actually hastened its loss. It turned out that it depends on a single species of ant, which carries the butterflies' pupae into its nests where they remain until the adults are ready to emerge. The ant only survives in short-grass pastures. Rabbits and sheep kept the grass just the right length for the ant, but the decline of rabbits and the removal of sheep in the mistaken effort to protect the butterflies' habitat allowed the grass to grow too long. The ant disappeared, followed rapidly by the butterfly. The programme that has seen the successful reintroduction of the butterfly is one of conservation's rare success stories. Achieving it required enormous effort, great political support, and significant investment. Everyone was delighted, me included. The purpose of the programme, however, beyond getting the butterfly back, is unclear. Certainly, it would be hard to explain in economic or existential terms.

This concern is articulated in the 'precautionary principle' which, in the absence of complete knowledge, demands that no action be taken that might be damaging. On the face of it, this principle seems sound. But in practice it creates paralysis. The managers of Mburo will never know enough to predict with complete certainty the implications of losing the brown-chested lapwing – or, conversely, what impacts cattle brought in to create lapwing habitat might have on other species. The managers of False Bay Park could not be sure what would happen to the suni if the nyala were left to their own devices, nor what the results of culling the nyala might be. At some point, decisions have to be made on the basis of whatever knowledge is available, even with the awareness that it may not be good enough. Science can help us reduce the risks of our actions, but it cannot remove them entirely. Unexpected consequences will happen.

Despite the uncertainties of predicting the future of such complex systems, conservationists have been sending out an urgent message to humanity. Protect every animal, every plant, every species, because if we don't the ecosystems on which we all depend may collapse 'For the want of a nail, the kingdom was lost' is reimagined as 'for the want of a snail, the forest was lost'. This apocalyptic message found resonance within and outside the conservation community. It had the power of all apocalyptic messages – the power to compel attention. The degree to which it represented reality became almost beside the point; the power was in the story: 'Beware! Act with care and prudence. Don't make a mistake you can't fix.' But like other apocalyptic messages, it did not lead to the change in behaviour we hoped for. Sinners will sin, it seems, despite the certainty of hell's reward; the rate at which we, the global community, emit carbon continues to rise despite certainty that the impacts of climate change will be serious. Ecosystems and biodiversity continue to be lost, though we well know how much we love or depend on them – or on some of them at least.

People do not respond well to messages of doom. We throw our hands in the air, shrug our shoulders, stick our heads in the sand. But putting this aside, there is something worrying about the apocalyptic message and the science behind it. There is enormous uncertainty about the behaviour of complex systems because our understanding of them remains imperfect. Almost everything I was taught as an undergraduate has been revised since. Ecosystems operate on almost infinite levels of complexity and scale, each level of detail revealing more details below it. Ecosystems are so sensitive to even tiny changes that they are ultimately unpredictable. Furthermore, when we intervene to save a species or increase or reduce the numbers of a second in order to manage a third, we make decisions without knowing the actual state or the direction of change of the population or habitat we are manipulating. We are shooting pretty much in the dark, with only our – good, hopefully – intentions to guide us.

When we decide to intervene in a conservation area, we take a position based on the biodiversity we find there in relation to the biodiversity we think should be there or would like to find there. How we decide what we expect to find is a mysterious process, and it is not clear how we separate expectations from desires, or a wish to find this species but not that one. Ecologists tell us that science answers these questions, but this is not so.

When I lived in False Bay Park, the nearby Hluhluwe Game Reserve[156] was managed under a plan that required it to be maintained as far as possible in the state in which the first white man to travel in the area had found it. I thought this an incredibly arbitrary way to establish the objectives of a protected area, but in truth it is little different from the way protected area objectives are generally set. This is so even when science purports to dictate not just how we must proceed but what we must conserve.

The first management plan for Mburo, prepared after long and exhaustive consultations, states the park's primary objective as 'to preserve and develop the values of the park by conserving biodiversity [and] maintaining ecological processes'. The plan goes on to elaborate that all the species of the park are to be preserved. Though too heavily focused on biodiversity for my liking now, it seems nonetheless a sound statement of purpose, firmly founded on scientific principles. But closer investigation reveals it to be as arbitrary as the plan relating to Hluhluwe.

As we saw in earlier chapters, Mburo's habitats and ecosystems have been through the mill, subject to serial impacts on massive scales over centuries. The Bahima's traditional management of the land kept the bush down by intensive grazing and trampling, periodic burning and bush-cutting. Colonial attempts to eradicate the tsetse fly destroyed the bush and decimated wildlife. In more recent years there was uncontrolled hunting, and periods of exceptionally heavy grazing by cattle. How the millennia of human use, variations in climate and natural disasters affected Mburo prior to the modern period I cannot say. What I know for sure, though, is that three species of mammal that were present in living memory are no longer present: giant forest hog, African hunting dog and roan antelope. They have all been lost since the 1950s. We also know that elephants used to pass through Mburo but do so no longer, and that although lions come and go they are largely absent.[157]

---

156 Though white South Africans are justifiably proud of the fact that Hluhluwe Game Reserve was Africa's first formal protected area, it is worth noting that Shaka Zulu had already preserved the area as a hunting reserve for himself and his chiefs and warriors.

157 Lions were a significant part of the Mburo ecosystem until cattle ranchers extinguished them. Though the Bahima had co-existed with lions for hundreds of years, the modern rancher apparently could not, and the last were shot in the 1970s. Nonetheless, there have been a least four occasions when lions returned to Mburo, probably making their way there from Rwanda's Akagera National Park, traversing a settled landscape for over 100 kilometres to find their way to Mburo. In each case, however, local farmers, anxious for their livestock, poisoned them.

The loss of these species, including the two predators at the top of the food chain and the elephant, which can modify habitats all by itself, can be expected to have significant ecological implications. The requirement to conserve 'all the species of the park' turns out to be a requirement to conserve just those species that happened to be in the park at the time the plan was written. Was the brown-chested lapwing present then? Other species might have arrived just as the plan was being drafted. It is not at all clear, then, how Mburo's required complement of biodiversity should be determined, or whether in fact it can be determined in any way other than by chance or by preference.

We have considered the simplest measure of diversity in nature, the presence or absence of a species. As part of my research, I wanted to assess the arguments for excluding cattle from Mburo based on presumed impacts on biodiversity. I decided to look at the impact of grazing on the park's pasture plants as a proxy for the impacts of grazing on the park's biodiversity and ecology.

I marked out transects that ran 100 metres through the park in places that I was confident had not seen cattle for a few years at least. I did the same in locations outside the park that I was equally sure had been heavily grazed for years. Every 10 metres along these transects I sat down with a team of botanists to identify and count every plant we found in a square-metre plot. We also estimated how much of the plot each species covered, and how much was bare soil. When crunched, the numbers certainly showed big differences between transects inside and outside the park, but not quite the differences that might have been expected.

Some plants were more common inside the park, and some more common outside. Some species inside, protected from the pressure of being grazed by cattle, had thrived and covered large areas at the expense of other species that they suppressed. Outside, the plants that the cattle did not like expanded at the expense of the species the cattle did like. This all went to show, unsurprisingly, that heavy grazing by cattle did indeed change the pastures, and in quite significant ways.

Despite these changes, though, every species we found inside the park we also found outside – and we even found a few extra ones outside. We were surprised, too, to find that plants classified as rare, both in the literature and in our counts, were actually more likely to be found *outside* the park. It seemed that heavy grazing by cattle, if it had any effect relevant to conservation at all, actually favoured biodiversity. Differences there definitely were – but no differences in overall biodiversity. So the magical measure we had used to describe the importance of Mburo couldn't, it turned out, help us decide how to deal with grazing by cattle.

So, if we are not talking about differences in biodiversity based on the presence or absence of species, what kind of differences *are* we talking about? If we talk about the relative abundance and dominance of species when trying to take a management decision – whether or not to allow cattle in, for example, then we must decide what we are trying to deliver. Do we want more or less of certain plants, and do we want them to be bigger or smaller? Are we trying to protect Mburo as it had been before the cattle arrived? That was centuries ago, and many things had been different then. Should we aim to recreate the Mburo that had existed even before people arrived? If these ambitions were too difficult, should we take a leaf from Hluhluwe's book, and aim to recreate the landscape that John Speke, the first white man to traverse the area, would have walked through had he drifted a little further west on his way to meet the King of Buganda?

None of these seem reasonable propositions. It seems more judicious to talk about returning the park to its state when it was declared a hunting reserve just 60 years before, or perhaps when it was made a nature reserve 30 years after that, or even when the National Park was declared just 10 years before the plan was written. These might be more achievable – but are they any less arbitrary? Pick any point in Mburo's long history, and you will have picked a set of species associated with the circumstances that prevailed at the time. Not only will those species be different from those of other times, but Mburo itself will be different too. It will look different, smell different, support different human activities, and be subject to interests that perceive its many values in different ways.

If we are trusting to science to tell us what to do rather than simply how to do it, the fluidity of history and the natural world means our decisions cannot but turn out to be arbitrary. This is because science cannot actually tell us what we should do; we have to decide that by ourselves. And if our decisions are to be based on what we want to protect, we must first look carefully at why we want to protect it in the first place. We must ask ourselves what we are actually trying to achieve. But next, before we get into a discussion of who the 'we' is in this, there is a need to consider more deeply how we decide what we are trying to achieve by protecting Mburo, for example, or indeed, any place.

If describing what we want to protect in terms of species turns out to be difficult, talking about conservation in terms of naturalness is even more so. It would seem that conservation science should be able to describe ecosystems in terms of whether they are more natural or less natural, but it can't. That's because naturalness, like beauty, is in the eye of the beholder.

Following my efforts at counting plants, I decided to assess how certain groups of people responded to the locations I had counted in. From the hundreds of photos I had taken of my vegetation transects I selected seven that represented the typical range of Mburo's habitats. Although they were all quite similar in terms of the species they supported, they looked very different; one, for example, was a tangled thicket of scrambling plants and vines, another was heavily cropped grassland, while another was tall grasses in an open valley.

I put these photos in front of three groups of subjects; conservation scientist colleagues, tourists, and Bahima pastoralists. I asked 20 from each group what they thought or felt about the places in the photos, and then to rank the places in order of preference. When they had finished, I asked them to explain on what basis they had placed one above another.

My conservation colleagues talked a great deal about naturalness. Many explained that they had chosen the places they believed were the most natural, the ones where the hand of humans seemed absent or minimal. They preferred those because they believed they would have more biodiversity. They were largely wrong on both counts. The landscapes they preferred were not actually more natural than the others, as all had been heavily impacted by people and their livestock. Nor, in fact, were the landscapes more diverse, at least not in terms of the pasture plants I had assessed.

The tourists and the pastoralists made surprisingly similar choices, and for surprisingly similar reasons. Both liked the more open places, feeling that they were safer. And both liked places they thought would be easy for them to move through – places that, in fact, had been most heavily grazed by cattle.

The selection by conservationists of landscapes they believed to be more natural was interesting. It seemed to fit the thinking that underpinned the early development of the national park ideal, but it fitted less well with the science of modern conservation that gives pre-eminence to biodiversity as the driver of conservation practice. The results of this exercise suggested that though science was instructing us to maximize biodiversity in our management, on the ground we continue to respond to that earlier holy grail of conservation, the idea of nature unsullied by the shuffle of human feet or the stomp of the hooves of their animals.

The search for naturalness is a tricky one because the concept itself is a tricky one. It is as easy to mistake what you are looking *at* as it is to mistake what you are looking *for*. The use of science to define what is in essence an idea, nothing more than a perspective, has us counting and measuring and comparing, but it cannot tell us at what point in an ecological process we are making our observations and thus in what context we are making our decisions.[158]

The characteristics we use to describe ecosystems are endlessly shuffled by natural events, storms, droughts and epidemics, and recurring cycles that operate over timeframes too long for us to observe or fully understand. Add to this the chaotic nature of complicated systems, where the flapping of the apocryphal wings of mythical butterflies can stir up typhoons, and we are all at sea. We don't know whether the biodiversity we find is what we would have found an eon before, a century before, or even just the year before. Nor can we tell what the presence or absence of something might signify or result in.

I experienced the vagaries of nature first-hand as a teenager on a school expedition

---

158 During a discussion I had with a room full of MSc students and their teachers in London over whether protected areas could be justified, a point of contention was the impact the creation of parks had on the systems they were established to protect. A professor argued that parks were fundamentally flawed because they could not be properly distinguished from surrounding lands in terms of naturalness. 'This,' he declared, gesturing around the room and through the grimy windows to the busy street below, 'is all natural.' I got his point though I could not quite agree.

to the northwest Highlands of Scotland. We were surveying the mice, voles and shrews living in the heather and grass-covered hills of Ben Eigh Nature Reserve overlooking Loch Marie, a landscape of beauty and grandeur.

We were instructed how to lay our boxy Longworth Live Traps, and despite carefully setting them on the runs and pathways of our quarries through the grass tussocks, and checking them morning and evening, our entire catch over two weeks was just four wee timorous beasties. We thought our failure must be because we were doing something wrong. The warden, our instructor and mentor, assured us that we were not, and imparted a valuable lesson. The lack of creatures falling into our traps did not mean they weren't there; it was simply that they weren't there *that year*. Those kinds of small mammals go through colossal swings in numbers between years, the fabled behaviour of lemmings being a product of this. We had simply found our mice and voles, or rather not found them, at the bottom of one of these cycles. They were still there, but in such low numbers that we failed to catch more than a handful.

When my colleagues insisted that the grazing of cattle in Mburo was unnatural and that the pastures would be changed, degrading the value of the park, their argument was as hard for me to refute as it was for them to demonstrate. But because so much of the ethos of conservation is based around this concept of naturalness, even though none of us can actually define it, I had to find something to counter their convictions. A well-rehearsed argument against livestock in parks in Africa is that the hooves of cattle, sheep and goats are harder than the hooves of Africa's wild ungulates, and that these foreign migrants from Asia[159] trample the pastures and compact the soils in ways that Africa's wild beasts do not. Goats in particular are considered the scourge of every living plant.

This may be correct, or at least partly correct – but everything, of course, depends on how things are managed. Great herds of cattle did trample wide tracks through Mburo, tracks that were very different from the meandering paths of antelopes. And as I found, heavy grazing favoured some plant species, the ones cattle didn't like and curbed the growth of ones they did. But these impacts can be reduced or even directed. Goats may eat every green thing if left to their own devices, but their numbers and their behaviour can be controlled, and their voracious appetites can be used to reduce bush where bush might not be wanted.[160]

159 The possibility that cattle were domesticated from wild aurochs in Africa, in the fertile and well-watered plains of what is now the arid Sahel, as well as in Asia, has already been mentioned.

160 Both Hluhluwe and Mburo suffer, if that is the right expression, from 'bushing over' – scrub acacia and other small trees creating impenetrable thickets. They might be good habitats for birds but are not great for plains game, tourists or pastoralists. Both reserves have at different times resorted to the mechanical cutting of bush, which is expensive and gives short-lived results. Goats might do a better job. Hluhluwe now has elephants again, which are likely to make a significant difference, but reintroducing elephants to Mburo is not likely any time soon.

How then do we think about naturalness and integrate the idea into the management of nature? Putting aside the potential contradiction in terms – is nature natural if managed? – the question is to some degree a technical one and ecological studies can point us in the right direction. But it is primarily a discussion that revolves around what we want to achieve, and that entails choices.

If we are concerned with biodiversity, should we want to increase it? Would it not be desirable to encourage new species into an area set aside for the purpose of protecting species? If new species find their way into our parks, why are we not pleased? Why don't we help them get established?

Uganda National Parks decided to bring a group of Rothschild's giraffes, taken from Murchison Falls, to Mburo. Different reasons were advanced for this. Those relating to tourism and to strengthening the conservation status of giraffes in Uganda by establishing a new population had merit, though both were challenged. The most vigorous challenge, however, was mounted on the grounds that giraffe had never lived in Mburo. Concerns were raised over whether they would flourish and how they might affect the park's ecosystems. But the strongest resistance came because it was considered wrong to disturb the natural composition of species by introducing a new one. The giraffes would make Mburo less natural.[161]

Wherever one stands on these questions, the way science and economics influence opinion about the very purpose of conservation needs to be thought through. At local and global levels, science and economics have shaped a perspective that makes conservation an existential imperative and demands that we all protect nature, locally and globally. This argument removes choice from the discussion, as there can be no choice when it comes to avoiding disaster.

In some ways, raising concerns about the survival of nature to this level of threat for humanity is undoubtedly right. If we poison the oceans there will be no fish in the seas and no fish to eat. If we erode the soil in a profligate manner we will be unable to grow food. If we blanket the planet with carbon dioxide and methane the world will heat up, and though we don't know exactly what will result from this we can be pretty sure it won't be good for us. And in the final analysis if all species are lost there will be, by logical extension, no life on earth.

These are real threats to our survival, certainly to the life we know and, with good reason, hold to be good. Averting these losses is within our capacity if we organize ourselves and take the right decisions. Regrettably, linking the avoidance of these disastrous scenarios to demands to protect nature is not entirely straightforward.

---

161 IUCN has strict guidelines for the introduction or reintroduction of species that firmly support the position that species should never be introduced into an area in which they did not formerly occur.

Proving that we need nature to survive is both easy and difficult. It is easy to show that without nature there is no life; we are a species, too, so if there is no nature there can be no humans either. Even arguments less extreme than this show that the loss of forests, for example, will reduce access to resources that people may need or want: water in the dry season or a suni antelope for the pot. Unfortunately, it is more difficult to argue that particular bits of nature – this species, that ecosystem, a particular place – are essential to our survival, either locally or globally.

Constructing arguments for protecting even the Critically Endangered primates I have worked with over the years is pretty difficult. During the seven years I spent in Vietnam, I was hard pushed to claim that the villagers were doomed if they failed to protect the Delacour's langur, the Tonkin snub-nosed monkey, the golden-headed langur. Even making local arguments for Uganda's iconic mountain gorilla, with its undeniable economic value, was not easy. Arguments for protecting species need to be couched in other terms. In the case of gorillas at least, with their close relationship to humans, they can be. It is more difficult for other species.

During my time in Vietnam I became responsible for a project to save the Cao vit or eastern black-crested gibbon, one of several species of apes and monkeys that are just about hanging on in Vietnam.[162] The project was designed when a group of 40 of these gibbons was rediscovered in 2000.[163] This is a tiny number, and the probability of such a small population surviving is low. A single storm, a bad drought or a new disease could wipe them out. Ignoring hard-headed arguments raised against getting involved, a team began work to bring them back from the brink of extinction.

We worked with the local community to stop their farming, hunting and wood cutting in the tiny area of forest the gibbons lived in, and we worked with local and national government to establish a protected area. What arguments did we raise to justify these actions? Nobody suggested that the forest the gibbons survived in, which no doubt harboured many other interesting and rare species, would wither away should the gibbons be finally and absolutely lost. That the gibbons might attract tourists was raised, but it was a long shot and tourism was unlikely to be important. Should the gibbons have finally winked out of existence, few would have noticed, and even fewer would have been impacted by their demise.

In the final analysis, the arguments for protecting the Cao vit gibbon were emotional and personal. There was no need to save it. It had been assumed extinct anyway. It was a choice. We wanted to save it because it was beautiful in our eyes. Its haunting song floated

---

162 Vietnam has the dubious honour of hosting 5 of the world's most endangered 25 primates, beaten only by Madagascar that hosts 6. Populations of these species number as low as 70 individuals and no more than 700. This is an indication of the degree of deforestation in Vietnam, which has reduced and fragmented the habitats of these primates, but also reflects the high levels of hunting of primates for food, prestige consumption and traditional medicine.

163 A second group was found a few years later, just across the border in China, raising the count to one hundred for the species.

through the mist as the sun rose. The local farmers also loved the sound of their calls and when they learned there were no more of these gibbons anywhere else in the world, they accepted responsibility for them, even giving them the name of their home-water.[164]

We and other organizations working in Vietnam made similar choices to save Tonkin snub-nosed monkeys, golden-headed langurs, Delacour's langurs, and grey-shanked douc langurs, all monkeys, but also giant Swinhoe's turtles, Owston's palm civet, striped rabbits and the saola, a wild ox only described by science in 1992, as well as many other species, including trees and plants. People interested in conservation, professionals or not, make the same kinds of choice when they decide to save a species, a wild place, a site of beauty, or merely something of value to them. These are choices, and we make them in response to what we hold to be important. We take these decisions not to avoid some global calamity, but because we consider the loss of an animal or a tree or a place to be a tragedy. We cannot always expect others to make the same choice, though we hope they will.

Once we admit that conservation is about making choices – often difficult ones – and complex trade-offs, and start to consider what it is that we want, we must begin to think in a more fundamental way about what nature is, both for us and for others. This will influence our reasons for protecting it. It will also require consideration of whose reasons are reflected. We must agree whose nature we are talking about, and who it is being conserved for. The big question concerning protected areas has, in truth, never been about *how* to manage them, though there are plenty of issues there too, but about what they are *for*.

We can probably agree that protected areas are for nature, but that does not take us far. We then have to agree what nature is and what nature is for. To answer these questions, we have to talk about values, the values of nature, the values we are trying to protect, the meanings of a tree or a place, what they mean for our identity, our creativity, our spirituality, and more. Once we have these on the table, the purported absolute values of economics and science that control the direction and practice of conservation must give way to discussions of what we hold to be important.

This is the point at which we have to figure out who this elusive 'we' is, and consider not just what I may want, but also what you may want, and what they may want. And then we must find a way to put all of our different values and perspectives and wants together and work together to protect all the values of nature held by all of us.

---

164 In Vietnam's two main river systems, people do not talk about their 'homeland' but their 'homewater'. When you ask someone what country – what land – they come from, you actually ask what water they come from.

7

Beautiful Beasts for the Beautiful Land

Shortly after I began work at Mburo, three new national parks were created in Uganda: Bwindi Impenetrable and Mgahinga in 1991 and Semliki in 1993. Together they separated the Batwa Indigenous People from the forests that had lain at the centre of their lives. Once the forests were declared national parks, the centuries-old connection between the Batwa and their forest was unceremoniously terminated. Tinkering with the fine lines of community participation and benefit-sharing did nothing to help. It can be argued these projects did little more than oil the wheels of a machine that continued to control access to nature and how it was understood, just as our work at Mburo did. We provided justifications for why we conserved nature and marginalized others' relationships with it.

Shocking though this sounds, at the time it was par for the course, and international donor agencies and wildlife charities – including my employer, the African Wildlife Foundation – were queuing up to support these new parks. Although I, at Mburo, was attempting to establish a slightly different approach, a marginally more open perspective, my efforts were still guided by the ideology of the exclusive protected area. Just as others were working to exclude the Batwa and others from their forests, Arthur, Moses and I were working to remove the farmers and herders from Mburo.

The wrong of much done in the name of conservation is self-evident. Ancient rights and connections between people and place were severed. Access to and use of land and resources were terminated. Communities were removed, often by force, from sometimes age-old homes. These acts were excused in the name of the global good they did, and explained locally as necessary to prevent the loss of nature's bounty, a loss that would fall most heavily on those most dependent on it. These were the rural poor, our neighbours. Even if it had been concern for poor local people that had driven the establishment of our parks – which of course it had not been – the claims to carry out conservation for the good of humanity in general and local people in particular while visiting injustices on these same communities were both morally wrong and wrong as a strategy.

Success and failure can look very different when assessed at different scales, and achievement may look like failure when examined in the future. I am sure that the warden who cleared Lake Mburo of its people in 1983 was pleased with his achievement, but the takeover of the park, the destruction of its infrastructure, the massacre of its wildlife, and the slashing reduction in its size that followed just three years later were direct results of what he would have considered a job well done.

The meticulously planned and executed relocation of farming households from the park that I had helped bring about was a success with respect to Mburo. Everyone left voluntarily in 1995, apparently content with the compensation we provided, and they did not return. But it is not clear that we achieved a net win for conservation. Some years after the event I learned that many of the families who took the offer of money and land to leave the park, sold their new land and decamped to a forest reserve 80 kilometres away where they settled, destroying much of the forest there.[165]

My 30-year association with Mburo gives me a particular perspective on the values of that beautiful place and how to protect them, but my experience has also changed how I think about conservation. The most conspicuous characteristic of the modern conservation movement has been its championing of exclusive protected areas. Though not by any means the sole form available to governments – and, as mentioned earlier, a range of forms exist, most of which do *not* demand the total exclusion of people and their activities – the exclusionary national park seems to be considered the best.[166]

Conservationists talk of elevating or upgrading protected areas of lower status to become national parks, clearly signalling how they are perceived. The most desirable seem to be the strictest, and as a result – by their very design if not by intention – those parks separate people from the natural world. Legislation establishing them and under which they operate emphasize long lists of forbidden activities, ensuring that most people and their interests are firmly excluded.

---

165 The process of relocating farming households from Mburo became a major undertaking. The agreement reached by the project, the donor, the Ranch Restructuring Board, the park and the households, was that the displaced people would be compensated for their fixed assets and given land. On signing an agreement to leave by a given date, money was deposited in a bank account opened for them. This money was supposed to be invested in the land that government allocated to them. In the event, many families took the project money, sold the land, and encroached on Sango Bay Forest Reserve. This can be thought of as a classic case of 'leakage'. In projects designed to reduce carbon dioxide emissions from deforestation – REDD+ projects – leakage refers to increases in carbon emission outside the area of the project that result from activities designed to protect forest inside the area. Leakage counteracts the reduction of carbon emissions in one place by increasing emissions in another outside the boundary of the carbon accounting system. The positive outcome for Mburo at the expense of Sango Bay was similar.

166 Not all protected areas that are called 'national parks' are the same, and not all are based on the exclusionary ideal of the national park derived from those first ones in the USA. In Great Britain, for example, the areas called National Parks do not exclude people, and indeed are largely formed from privately owned and farmed land.

Of course, not everyone is kept out. Many parks welcome tourists with enthusiasm, along with the enterprises that cater for them. Given the rhetoric surrounding the parks, a surprising range of activities and facilities are allowed, including hotels, helicopter pads, golf courses, karaoke bars, swimming pools, scuba diving, hot air ballooning, quad biking, fishing – the list goes on. Park officials generally live inside, often with their families. Those providing the many services that 21st-century conservation depends on – mechanics, builders and technicians of all kinds – also tend to be located inside the parks.[167] Scientists are welcomed, too – indeed are considered essential members of park institutions.

Some protected areas are considered so pristine and so valuable or vulnerable that the separation imposed between people and nature is more democratic, to the extent that everyone is barred entry. More generally, though – certainly in most of the parks that I have spent time in – it is primarily the local people whose presence and activities are excluded. Except for occasional bridge-building visits organized for local leaders and schoolchildren, the park's wardens and rangers work hard to ensure their absence.

Separating people from nature in order to conserve it, especially when separation is primarily imposed on local people, seems to make little sense. The attention paid by conservation projects to environmental education suggests that we know very well that parks need support to succeed in the long run, and that ultimately policing, fines and fences will fail. Efforts to build support through education seem to be based on two assumptions: that support for conservation requires people to understand the importance of nature to their survival; and that the values conservationists perceive in nature, whether beauty, diversity, intrinsic or economic, must be accepted. Our self-generated need to go out and educate people suggests that we do not believe, or perhaps have not considered, that our neighbours might have their own understanding of the natural world and their own perceptions of its worth. It also suggests that we have not understood that this knowledge, these beliefs, might be important in designing not just education programmes but the protected areas themselves.

It is an irony, which still has the power to surprise me, that despite believing that we need the support of our neighbours, and given all the activities we undertake to strengthen this, we continue to demand the exclusion of these same communities from the natural world that we want their help to protect. Whatever we think we are achieving

---

167 I have noticed that though protected area managers work hard to keep people out of their parks, and even tolerate tourists and the infrastructure of hotels and camps somewhat grudgingly, they like to locate their headquarters plumb in the middle of the most remote, beautiful, interesting and no-doubt sensitive locations they are responsible for protecting. Arguments of strategic placement for reasons of ensuring protection and security are weak and it is pretty clear to me that sites are selected just because this is where the managers want to be, deep in the heart of nature and far from the madding crowd.

in the short term, placing barriers between people and nature must surely be a mistake. It creates a downward spiral of negative feedback in which weakened connections to nature resulting from separation leads to reduced interest in it, resulting in less support for its conservation. Arguments designed to convince ordinary people and their governments that protecting nature makes economic sense seem equally dangerous.

From a personal perspective, both these positions contradict my own interest in the natural world. Financial incentives to protect nature are beside the point for me, because I love that world; and connection to nature, the chance to engage with it closely and intimately, is my reward for protecting it.

It is hard to reconcile but true, nonetheless, that though we assiduously promulgate economic arguments to justify our demands to protect nature, conservationists are not generally motivated by these arguments. It seems that our insistence that market forces and money will protect nature are unrelated to our own, often fierce, commitment to its conservation. The perspective we give to governments, communities, politicians and business leaders are not those that have led us to devote our ourselves to nature. We talk about the financial importance of nature and insist that only practical arguments based on financial outcomes will sway decision-makers to take the steps necessary to conserve it.

Although our own decisions and sacrifices are motivated by love, not lucre, we seem unable to believe that others might be the same. It is a mystery to me how this contradiction has persisted so long – yet it has.

When, halfway through my research, I discovered the concept of enyemibwa, that was what taught me that the Bahima were as a people culturally driven by the desire to increase the beauty of their herds. This forced me to set aside my conditioned conception of Mburo and rethink what was important about it. That the Bahima had for centuries called their land the Beautiful Land made clear to me that they had a relationship with Mburo that owed nothing to my notions of land, beauty and wilderness. I also saw that their values in the land – what it meant to them – lay at the heart of their struggles against the park and officialdom in general. This emerging understanding gave me a new and compelling explanation for the conflict I had spent years observing and trying to moderate. It then led me to challenge most of what I thought was important about conserving nature, and to question many of the actions I had carried out in its name.

I had experienced Mburo as a panorama, a canvas adorned with plants and creatures draped across a scaffold of terrain, now burned by the sun, now drenched by storms. I saw its elements fashioned through the ages and the unfathomable exchanges between them. I saw Mburo as a treasure, and I wanted to place it beyond unbounded human demands. But now I can glimpse aspects of a different landscape, one constructed in the minds of

Bahima over the centuries of their occupation. I might see this land as beautiful, but its meanings are essentially theirs, not mine.

In the most southerly part of the park there is a hill called Mujwiguru. I drove there when I could, threading the acacia trees and ant hills along the track that Moses had opened. Alternating scents of crushed grass and red dust would drift through my car window, I would catch the flash of scarlet as a turaco hurtles with a clap through tangled branches and glimpse the hulking form of a buffalo in the vegetation growing along the edge of the lake.

From its modest summit the hill commands a 360-degree view over hills, valleys, lakes and swamps. The land dissolves in a yellow haze in the dry season or is etched in detail against heavy skies when rains wash the air clean. Mujwiguru translates as something like 'where the heavens mingle with the earth'. I liked to sit there, in the hope of feeling an experience in common with the Bahima, but the name is more than a description of the expansive view. It is suffused with meanings beyond my comprehension.

The hill is an ancient site of worship. Potshards and half-burned lumps of charcoal, old and not so old, were scattered across the hill. The Bahima prayed there to their ancestors and the gods who once walked the earth, mingling with them as other gods have done in other lands. Mujwiguru was a place where those ancient connections between heaven and earth, between gods and mortals, could be remembered and celebrated. It was not a place simply to feel the peace and tranquillity of nature, but a reminder of a time when the gods and people lived together in the Beautiful Land.

Every hill and valley there has a name that refers to times and events from the histories of the Kingdom of Nkore and before.[168] Swamps collect in furrows carved by an ancestor's spear thrusts. Streams burst from the ground where it was pierced by arrows. Some places are remembered in recitations as the sites of battles where heroes fell, or cattle were raided.

Left: The beauty of Mburo through my eyes
Right: The beauty of Mburo through a Muhima's eyes, tall grasses in open valleys

---

168 Oral history does not make a hard separation between these two, with the first king of Nkore, Ruhinda, being the son of Ndahura, the last of the Bachwezi, the god-like ancestors of the Bahima.

Left: *The Kigarama Hills seen west of Lake Mburo*

Upper right: *The beauty of Mburo through a Muhima's eyes, good grazing open for the cattle to move through safely*

Lower right: *The view from Mujwiguru Hill*

Enyonza, Carissa edulis, *berries were collected for food during famines and during the social stress that followed the rinderpest epidemic; they were also used as bride price instead of cattle*

The trees and shrubs and flowers are part of the story too. *Omugari*, the camel's foot tree (*Buhinia veriagata*), is praised for its leathery seed pods that feed cattle when grazing is sparse. Its branches curve down to the ground, giving deep shade during the day and warmth and protection at night; herders grazing their cattle far from home would bivouac in their embrace. *Enyonza*, an unassuming spiny shrub, carries creamy white flowers and bright red berries. It holds a favoured place in songs and stories as the saviour of the Bahima.[169] It was impossible to persist in defining Mburo as a wild place for wildlife when it was so clearly a place of the Bahima, their cattle and their ancestors.

When I began the Community Project, I was excited because I felt I was helping to undo the troubles of Mburo's past; I was going to build bridges based on common interests and understandings. But instead I set about attempting to instill a conservation ethic founded on my values.

The project followed the well-trodden path of educating and building awareness. The adults were already beyond us, their prejudices fixed; the children, however, could still be convinced, and might carry messages back to their elder brothers and sisters, even to their parents. Exposing children to the wonders of nature and explaining the intricate interactions between the park and their future wellbeing would make conservationists of them, wouldn't it.

When it came to education we were pushing at an open door. Parents believed that education would be the saviour of their children, and that they would grow up to be doctors and lawyers and government officers in suits, driving shiny cars. Not for *them* a life of grubbing in the soil or toiling with the cattle, at the mercy of the weather and the wildlife, the hard life that had been theirs.

So we worked with communities and local leaders to put roofs on classrooms, build latrine blocks and provide housing for the teachers. We fielded volunteers from America and Britain to write textbooks, explain ecology and establish wildlife clubs. We built a fine education centre and equipped a lorry as a bus to bring students into the park. For the adults we organized open days, and we reached out to community elders and leaders, busing them into the park to see what we were doing there.

Meanwhile, theorists and policymakers were working to conceive the worth of nature

---

169 The berries of the enyonza bush (*Carissa edulis*) have a deep red colour when ripe, while their flowers are a creamy white. All parts of the bush release a white milky sap if cut. The deep red and the white represent the perfect combination of hide and horn in the idealized Enyemibwa, and lie at the centre of the Bahima aesthetic. Enyonza appears in songs and recitations describing the cattle in the landscape but also has historical significance. During the rinderpest epidemic that destroyed 90 per cent of the cattle of Nkore in the late 19th century, the berries were collected. Not only were they eaten to stave off hunger and famine but baskets of them were used as bride price, allowing marriage ceremonies to continue even though the cattle normally exchanged were unavailable.

Moses explains the African fish eagle's significance for
Mburo at the opening of the Education Centre

in material terms.[170] As market fundamentalism swept all before it, the west engaged with developing countries almost entirely in economic terms. Conservation was not immune, and both donors and recipient governments demanded that conservation had to be economically sustainable, meaning that it had to pay for itself.[171] It was not unreasonable to think that our poor neighbours might not agree that Mburo should be protected for its own sake, but the idea of convincing them it would pay for itself was hopelessly optimistic. Nonetheless, our programme, just like the others, focused on the role that nature played in bringing rain, bringing jobs, bringing income, bringing schools and clinics, and all the rest.

Though we focused on making the case for the park in these terms, we did not ignore the other values of nature entirely. We worked in Mburo because of our love of nature,

---

170 Margaret Thatcher, the British prime minister of the 1980s, did more than perhaps any other political leader to espouse the political ideology of neo-liberalism advocated by the New Right, which was central to the political philosophy often referred to as Thatcherism in Britain or Reaganomics in America. Thatcher had an advisor for this very purpose, a specialist in contingent valuations and other arcane methods for expressing intangible values in terms of financial worth.

171 When trying to represent my concerns over this proposal, which sounded fine on the surface but was deeply problematic, education was a good analogy. Almost everyone agreed that education was one of the fundamental platforms required for achieving economic development, not to mention central to equality and to quality of life. Most governments accept that education had to be invested in. Schools and teachers have to be paid for and cannot be expected to turn a profit or pay their way. They are not in those terms, sustainable. Nature conservation is the same. Conservation does not need to turn a profit in narrow financial terms in order to be understood as a key part of a sustainable nation or community.

A young topi suckles in the shade of Acacia gerrardii *trees flushing green with the early rains despite their ragged trunks*

and we could not avoid praising its beauty and its wildlife or putting forward ethical and even moral arguments for protecting it. When not in Full Economic Display Mode we insisted that the children – and *their* children, in turn – should enjoy nature as we did; impala standing in dappled morning light, deep in the yellow grasses, flashing pied ears and snorting; the harsh cough and rasp of a leopard sawing in the inky darkness of a stormy night. We argued that the smell of acacia pollen, flooding across the hills and valleys as the seasons changed, one species following the next was their right, too – the yellow polka-dot flowers of scraggy *Acacia hocki* standing in thick clumps; the ivory flowers of *Acacia polycantha*, like tallow candles; the creamy clumps of heavily armoured *Acacia sibiriana*; and the sparse white flowers of *Acacia gerrardii*, whose twisted trunks and branches festooned with lichen make even young trees look old.

This desire to stir a love of nature in the hearts and minds of our neighbours was not wrong, but it was, unfortunately, based on the assumption that these people did not love their natural world already. How presumptuous of us to tell people, whose links to Mburo went back hundreds of years, how special it was. I count myself fortunate that, sitting in the shade of trees, in schoolrooms, in *kafundas* drinking waragi and banana beer, I was not laughed out of court when I described the wonders of Kaaro Karungi, the Beautiful Land.

Perhaps they were suspicious, fearing I would contrive to steal away anything they told me was important. It had happened before. Perhaps they did not habitually reflect or think in terms of values; they were just there and always had been. Or perhaps they had been so well trained by our visits that they fed back to us our own stories of the Madagascar periwinkle, rainmaking or how tourism would make them rich.

Rather than telling them how important zebras were by grazing down the long grasses to provide opportunities for the delicate oribi to feed, or explaining that hyenas kept the wild herds of animals healthy – all of which I am sure they knew already – I should have been asking questions. What did they think was important about Mburo? What did they love about it? What would they protect if given the chance? If I had started out asking questions, rather than telling our neighbours what and how to think, I might have had more success in gaining their support. I would certainly have discovered a lot earlier than I did that they had profound cultural and personal connections to the land and to nature, and I might have recognized sooner the role that culture should play in protecting nature.

Sometime in 1996, halfway through the second phase of our project, when we were engaging with the newly established Uganda Wildlife Authority to create the Community Conservation Unit, I was called to a meeting in Nairobi to discuss the direction the African Wildlife Foundation might take. Our president, the highly respected conservationist Michael Wright, chaired a session to unearth what he called 'big ideas' for conservation.

I had already begun my doctoral research and was learning to discuss in general terms why, when we thought about conservation, we also needed to think about culture. My work with the Bahima to understand their connections to nature was helping me glimpse the world as they saw it. This helped me understand their conflict with Mburo as a park, and might identify synergies between their values and the values of conservation. I was learning to articulate this idea as a guide for a cultural approach to conservation. I decided to try it out at the meeting.

'The modern conservation movement is failing to build support for our protected areas,' I suggested at an appropriate moment, 'because we describe what we are doing in the terms of science and justify it in terms of economics. But in most cases, this thinking doesn't mean much to our neighbours. It lacks resonance. But if we think about conservation in terms of local values, and learn to describe it in the language that our neighbours use about nature, surely this would be a more effective and enduring basis for building partnerships? If we look at conservation only through our *own* values and describe nature in these terms, it's not surprising that most people aren't interested.'

My colleagues were not convinced. They resisted the suggestion that science, the disinterested pursuit of the true nature of nature, should be put on a level with the myths and emotions of the 'uneducated' and 'superstitious' villagers. They opposed the idea that local people could hold any sets of values, whether their own or imported from the west, above the demands of their immediate material needs. Poor people did not have the luxury of valuing nature for its beauty or for other intangible or ethical reasons. The communities we worked with were focused on survival, and unless they accepted that nature contributed to that, they would never protect it.

My point was that to build synergies we don't all need to have the same values or see nature as important in the same way. Partnerships based on mutual economic interests might work sometimes, but in most cases they don't. The hope of imbuing our western way of loving nature, based as it is on a particular set of values, would be a long shot, and in any case is a form of cultural imperialism.

But validating values held by people of other cultures and aligning them with our own could be the basis of a partnership in which we pursued different interests through the same outcomes. A partnership to protect Mburo could be founded on our interest in conserving buffalo or oribi combined with the Bahima's interest in conserving Ankole cattle in the same place at the same time. A single park could express multiple layers of value simultaneously if managed in order to do so, and if compromises were made on both sides. The biggest compromise for conservationists would be letting go of our conviction in our absolute values and accepting a more relative view of the world. There would also need to be more practical compromises to allow multiple values to be accommodated.

The park managers at Mburo would have to accept the presence of cattle, and the Bahima would have to accept the place of wildlife. Neither of these stipulations should represent serious difficulties, though others would be more difficult, and some might be near-impossible. A community might never accept giving up their ceremonies of respect to their ancestors, for example, and conservationists might never accept the sustainable use of elephants or gorillas.

The sustainable use of wildlife has been presented as a means by which conservation can be achieved, especially in southern Africa. Using wildlife, mainly but not only, by hunting it or cropping it, generates revenues and thus reasons for conserving it. The principle is hotly contested, though, especially for some species.[172] And some areas of conflict might be best set aside, because focusing on them could prevent potentially successful compromises from even being discussed.[173]

I did not win my colleagues over, but I did make some progress, and Michael put my idea on his list of 'big ideas'. But as things turned out, nothing much happened as a result of my big idea. I was disappointed but not surprised. I concentrated on my research,

---

172 Some of the most intractable conservation problems seem to be firmly based on cultural differences on what is acceptable human behaviour. The killing of whales is a prime example of this; it caused half a century of paralysis within the International Whaling Commission. The global majority insist that it is immoral to harvest whales, and attempt to impose this view on the small minority who insist that hunting whales is central to their culture.

173 The creation of the Gwaii Haanas National Park Reserve on the west coast of Canada came about despite the fact that the Haida Nation were pursuing a legal claim of ownership against the Canadian government. An agreement was signed that placed the management of the reserve in the hands of a Board comprised of equal numbers of Haida and government representatives. The Board must reach decisions unanimously, forcing compromises, and is not permitted to discuss the ownership of the land, which is left firmly off the table.

while following the direction many conservation charities were taking, the organization concentrated on wildlife industries.

In 2002, having completed my thesis, I accepted a job with Fauna & Flora International, a British conservation charity. Then, when living and working in Vietnam, I learned that Michael Wright – he who had styled my culture in conservation idea a 'big idea' – had joined the MacArthur Foundation, a major international funder of conservation. This was an opportunity.

Arthur had joined Fauna & Flora shortly after me, and together we developed the Culture, Values and Conservation Project. From 2005, for the next ten years, the MacArthur Foundation funded it. Throughout this period Arthur managed three phases of the project, while I made periodic trips to Uganda to support him.

In 2007, on one of my precious visits, I found myself back in Sanga. Little had changed in the scruffy trading centre where I had turned off the tarmac road in 1988, almost two decades before, in search of the reserve that lay somewhere beyond. Now, 20 or so of us sat in the offices of the sub-county administration, newly built, smelling of wet cement and whitewash, and filled with the gentle scuffling and musty smell of the bats that had already discovered its roof space. The meeting brought together Bahima leaders, renowned breeders of Enyemibwa, with park and government authorities, and the project team. We were there to discuss how the values provided by the Ankole cattle could be integrated into the management of Lake Mburo National Park.

My research had suggested that the Bahima's pastoral worldview, rooted in a special connection to the Mburo landscape, would, if incorporated into the park's management, deliver a wide range of positive outcomes for both the park and the Bahima. It would broaden the values for which the park was recognized, give it local resonance, remove a source of perennial conflict, and help preserve the Bahima's cattle and culture. The project, developed with the Uganda National Parks and implemented in partnership with them, had been established to deliver on this, and now detailed discussion with all parties was needed to see what this meant in practical terms.

After prayers, we talked around the implications of the Bahima adopting a settled life on their plots of land and giving up the nomadic life, as demanded by President Museveni. The survival of their long-horned cattle was in doubt, which meant that the practices of breeding and keeping Beautiful Beasts were also threatened. Though many Bahima declared they would never give up their special cattle, for most of them the fragmentation of the land, the economic realities facing them, and the arrival of exotic bulls in the herds, all foreshadowed the loss of the Ankole cattle. We easily agreed that it was important to conserve it as a breed, and that the Beautiful Beasts were something special to Uganda as well as to the Bahima. It was also accepted that concerted actions would be needed to save the long-horned cattle.

Differences of perspective and priority were of course evident. The park officials were concerned about the future of the park and its wildlife. They pointed out that the incursions of so many bovines made it hard for them; that chemical acaricides were killing oxpeckers; and that the rampant hunting going on outside the park was steadily reducing the animals inside it. More income was needed to maintain the roads, pay staff and share revenues, and *that* meant attracting investors and increasing the number of tourists – which was of course impossible with so many cattle and so little wildlife.

Although everyone agreed that the park was important, that many values that were part of Ankole would be lost if the park was lost, and that everyone should support its protection in the future, it was hard to keep the conversation on track and positive. Some Bahima, who kept larger herds than they could support on their own land, thought they should be allowed to graze in the park. They raised the old arguments about shortages of water and the need for milk for their families, and reminded us of the ancient links between Mburo and the king's royal herds. The authorities countered, as ever, that the law was clear and could not be changed, that the cattle destroyed the grazing and outcompeted the wildlife; that the park was for wildlife and tourists. We were still locked into the oppositional positions born from conflicting cultural values.

Suggesting that participants refrain from taking positions, for the moment at least, we moved the discussion back to what the parties had agreed and committed themselves to achieving together. Solutions for difficulties could be found, as long as both sides could make compromises.

Just as Mburo belonged to the nation rather than individuals, and was protected for the benefit of all, the conservation of the long-horned cattle might be achieved, I suggested, if the animals did not belong to individuals but were conserved for all. It was hard for Bahima to conceive of cattle which individuals did not own, but after some discussion it was accepted that a herd established to protect the breed could exist outside individual interests. Such a herd would also help conserve the knowledge, practices and beliefs associated with breeding, caring for, praising and naming the Enyemibwa. These animals would be 'cultural' cattle and belong to all the people, rather than the 'economic' cattle, which naturally belonged to individuals.

With the concept of cultural cattle on the table, the park officers agreed that accepting those animals in the park was a prospect quite different from that of accepting cattle in the park for economic reasons. The animals' presence would not be justified in terms of production, but in terms of conservation – not of wildlife, true, but conservation nonetheless. The Ankole cattle contained unique biodiversity, and they had national and cultural significance.

Setting an economic precedent that would compromise the park's position was thus avoided. The concern that if 100 animals was considered a good thing then 1,000 or 10,000 would be even better would not apply to a cultural herd. The discussion around 'how many animals' would be in relation only to the number needed to achieve conservation objectives, unmuddied by financial considerations.

The proposal to link the conservation of land, wildlife, cattle and culture evolved into a wider discussion in which the protection of different values such as biodiversity, ecological functions, naturalness, historical value, ethnic identity and connection to place could be balanced. Now we could begin to talk about how the integration of the long-horned cattle could strengthen the park.

If the park became the home to the most beautiful beasts in Ankole, as it had been in the days of the king, there would be many who would come to see them, and not just Bahima. For the most part these people might have little interest in wildlife, but they would come to see the enyemibwa. And if the park became known as the home of the Enyemibwa, the warden would not have to rely on values that people did not understand, believe in or care about to justify the park, but could instead talk of the Beautiful Beasts and the Beautiful Land, and keep his thoughts about the beautiful wild animals to himself.

It might also help resist the incursions of the economic cattle that he was still powerless to prevent, allowing him to argue that the park should not accept the cattle of individuals, as this would be unfair. Further, it would now become clear that those who pushed their herds across their neighbours' lands threatened the efforts to conserve the long horns as well as the park, damaging everyone else's interests to satisfy their own. Many of those who pushed their cattle into the park were well-connected army officers or government officials, and their cattle were stunted cross-bred creatures with black horns. As national parks were for the nation, Mburo, by resisting those incursions, would help conserve the nation's cattle for future generations, just as it did its wildlife.

As we felt our way into this dialogue, it became clear that we were describing the very different kinds of benefit that Mburo could offer to the different parties. I began to sense their excitement. If part of the idea was to conserve the knowledge and practices associated with keeping the Ankole cattle, as well as the breed itself, there were many things to consider.

The essence of the animal could not be separated from the practices by which it was bred and cared for, or the place in which it existed. The bovine, with its long legs, was bred to walk. This is why Bahima resisted fences in the Beautiful Land and fought against its division. Breaking up the land changed both the place and the cattle, subtly altering their values and meanings. When it rained in the distance, the cattle would lift their heads, snuff the air, listen to the grumble of thunder and begin to walk, heading into the wind. Outside the park this was now becoming impossible for fear of fear of falling into pits, trampling crops or tangling with wire fences, but inside the park the cattle could be allowed to walk freely.

The warden was concerned about the spraying of cattle with the chemicals that killed ticks, because that then poisoned the oxpeckers, the raucous birds that scamper around on the backs and sides of animals, feeding off those ticks. A Bahima elder agreed that Ankole cattle should not be sprayed – but for different reasons. He wanted to see them looked after properly, as they had been when each animal was carefully inspected and the ticks picked off by hand. Then in the park, when the pastures became infested with ticks and the cattle became sick, they should be driven to a new location, and once the pastures had grown tall again someone would return to burn them. This would kill the ticks and help keep the pastures free of bush.

Another elder, warming to the theme, suggested that the cattle could not truly be considered Ankole cattle unless a few at least were eaten by lions – and also that Bahima would not be true Bahima unless they knew how to defend their beasts. Not all would go that far, but they did agree that young people would benefit from learning how to build enclosures to protect the cattle and sitting with them at night as their grandfathers had done.

Though the meeting had started with the participants entrenched in positions reflecting the decades-old conflict, the opportunity to talk about the park in terms of values and to hear people talking from different perspectives helped move things forward. The participants were able to acknowledge that although their values were different they were not necessarily contradictory, and this opened them all to new ways of thinking. They saw what the park meant to others and how compromises could result in new values and benefits being generated. If new ways of managing the park could be envisaged and agreed, new partnerships could be established.

As the day wore on, we came back to the question of whose cattle these cultural cattle should be. There was some sense to the suggestion that the park itself should own them; this would certainly make the links abundantly clear between the conservation of wildlife, the landscape and the cattle. It would also facilitate the management of the herds in the context of a protected area that had to preserve multiple values and fulfil many functions. The Bahima elders, however, were not quite comfortable with this suggestion. It had been hard enough for them to accept that the link between Ankole cattle and individuals could be broken, and it was too hard for them to accept that cultural cattle, kept to protect the values of the Bahima, should be owned by the park.

The ownership of the Enyemibwa gave the Bahima prestige, both as individuals and as a people; these were the Ankole cattle of the people of Ankole. Further, it was the exchange of cattle that established relationships and friendships that persisted over generations. That an individual's standing in the community was based on their cattle guaranteed that the animals would be properly bred and cared for. The Bahima were not convinced that the park authorities could meet these demands, nor indeed that the park, an institution that had ridden roughshod over their values for decades, could be entrusted to care for them now. They would not abdicate their responsibilities to the cattle, to the land and to their ancestors.

An association of those committed to conserving the Enyemibwa culture was needed. Individuals would give cattle to this association, which would work with the warden to manage the herd within the park. Together they would decide where the herd would be grazed, when the herd needed to be split, and when there were enough animals to meet their objectives. The association would ensure that the cattle were properly cared for, ensure they were bred to be beautiful, and pass on knowledge to the next generation.

Thus it was that in 2009 the Ankole Cow Conservation Association came into being, and some time later the Founding Herd was established, 19 members each giving an animal.[174] The formation of the association marked the beginning of a journey for its members and the managers of the park, a journey that still continues.

This was an exciting meeting for me. While the long day passed outside, with the sounds of goats skittering by on the packed earth, the chatter of schoolchildren scattering homewards and the impossible roar of rain on the iron roof sheets interrupting the proceedings, there was a focused energy inside the room. I had sat in many meetings with local leaders to discuss ideas for activities that would bring benefits to their communities, or resolve interminable conflicts and complaints, or explain the values of a park – but at *this* meeting there was a degree of engagement, an enthusiasm, an openness and inventiveness amongst the participants that felt different.

There was no question about the interest amongst Bahima in linking the protection of their Enyemibwa to the conservation of the last remaining area that retained elements of the Beautiful Land. The wardens and rangers, though wary, saw the benefits of such an arrangement.

Sadly, despite struggling for ten years to negotiate agreements between the Ankole Cow Conservation Association and Mburo management, building up the Founding Herd, and establishing the Enyemibwa Culture and Education Centre in the park, we failed in the end to turn the potential of the cultural approaches into practice.

It was recognised that a cultural values approach would help Mburo's managers reduce conflict with the Bahima and reduce uncontrolled grazing in the park. It was accepted that integrating cultural values into the park would generate local support and add a visitor attraction. But the institutional resistance to cattle in the park was too strong. The absurd outcome was that the centre built to showcase the Enyemibwa was located inside the park – but the animals themselves were not.

In practice, however, the herd spent most of its time grazing the park, and a dam was even constructed for them. But the enclosures where they spent the night were built just outside the boundary, and to see them being milked the visitors to the Enyemibwa Centre had to leave the park. This unsatisfactory arrangement shows that although the authorities

---

174 It might not immediately seem such a big thing for a pastoralist to give a cow away, even in this new context where there was no assurance they would get one back in the future. But the gift of a cow is always a generous one. If a family need 100 cows to be economically comfortable, giving one away is equivalent to giving away 1 per cent of their wealth. There are many reasons for giving cows, though, and social structures through which the act of giving or even lending cows delivered broad social benefits. Nonetheless, the act of giving a cow was, and remains, a significant one.

The Founding Herd started with donations of a bull and some heifers and calves

Construction of the Enyemibwa Centre starts with the Demonstration Hut

saw the value of integrating local culture into Mburo they could not overcome their prejudices against the cattle. Clearly, efforts to gain true support for and understanding of cultural approaches had a way to go.

8

Sacred Mountain, Father Forest

When the MacArthur Foundation agreed to support our Culture, Values and Conservation Project in 2005, they insisted that we also work in the Rwenzori Mountains National Park, basing our approach there on the practices developed at Mburo as the foundation's programme was focused on the biodiverse forests of the Western Rift Valley, not the Ankole savannahs.[175] Nevertheless, the work in the Rwenzoris probably saved our project, because we showed that the cultural approach could deliver results quite quickly. At Mburo progress had been slow and problematic because of the deep roots to the conflict over the cattle, but the comparatively fast and easy progress in the mountains helped build impetus for the approach within the Uganda Wildlife Authority.

'At last!' exclaimed an elder during one of the early meetings to explore cultural relations between the Rwenzori Mountains and the community, 'After all these years, and all those meetings to talk about the park, *finally* you are talking about things that are important to us.'

In 1888 Stanley had marched right past the Rwenzori Mountains without, apparently, seeing them. Though their highest peak rises to 5,109 metres (16,762 ft), the third highest in Africa, and it is topped by permanent snow and glaciers, the mountains are often hidden in clouds and haze. The Rwenzori Mountains are home to the Bakonjo and Baamba peoples, who for centuries have farmed the steep foothills, harvested resources in the forests, and hunted in the highland moors.

The Banyarwenzururu, the people of Rwenzori, are keen to mark their independence from the lowland tribes. Living at high altitudes, it seems, emphasises their separation and reinforces a sense of difference. The Banyarwenzururu probably accepted the British administrators' decision to integrate them within the Kingdom of Toro in the early 1900s because it made little practical difference to them, secure as they were in their high

---

175 Donors to conservation, including private foundations like the MacArthur Foundation as well as government funders, often limit their support to specific locations. This may be because of the high levels of biodiversity of sites – so called hotspots – but it can make it difficult for government agencies and conservation charities to pursue their own priorities and programmes.

Left: Giant Afroalpine vegetation thrives in the mountain heights that are amongst the wettest in the world

Below: Smoke and dust conceal the peaks of the Rwenzori Mountains during the dry seasons

mountain villages. But on the eve of Uganda's independence in 1962 the Kingdom of Rwenzururu was declared. The new Ugandan government took little time in suppressing it, but a low-level insurgency continued for half a century until in 2009 the king, the Omusinga, was recognised by Yoweri Museveni's government.

The struggle of the Banyarwenzururu for recognition is best understood in the context of the mountains themselves. As noted elsewhere around the world, mountains hold great spiritual power for people living on their slopes.[176] Life at high altitudes propagates a

---

176 Examples of connections between mountains and spirituality have been long recognized, whether amongst the Buddhists of Tibet or the Inca of Peru. The Mountain Institute (http://mountain.org) is perhaps the most prominent organization that promotes the conservation of mountain landscapes and the spiritual values they promote, and has been an important focus for supporting the conservation of mountains and their many values around the world.

spirituality that appears to run stronger and deeper than for those living more easily in the lowlands. It seems embedded in the physical terrain. Perhaps the incessant demands of ascending and descending the precipitous paths, getting the blood pounding through the veins, inspires visions. Perhaps the dark storms that rush across the plains below to strike the exposed villages, or the clouds that roll down from the peaks to engulf them, contrive to inspire a sense of permanent awe. Or perhaps it is simply the proximity to the heavens.

Whatever the reason, the Banyarwenzururu have a strong awareness of the sacred, and although most are Christian today the majority retain a connection to the ancient gods. As the gods live in the mountains, the people have an appreciation of living closely with them. This reinforces the keenness of their connection to the mountains and has helped their traditions to persist. They believe that the power of Nyamuhanga, the Great Creator, flows through the gods and spirits to Omusinga, the king – and through him to the chiefs and headmen who govern the communities living on the great mountain ridges that descend to the plains.

We did not have a body of research on the relations and interactions between communities and the Rwenzori National Park as we had had for Mburo, but there was no shortage of knowledge about the Banyarwenzururu and their beliefs; the sacred nature of the mountains was well known. Nonetheless, when the National Park was created in 1991, and despite the intensive consultations that went into its design and planning, no mention was made of these relationships. Even when the park was inscribed as a World Heritage Site three years later, these relationships were ignored.

Sadly, these omissions are not surprising. The authorities considered the mountains and forests largely as had the British administrators when they had declared the Rwenzori Forest Reserve in 1941, with the timber supply removed and biodiversity added for lustre. The forests were to be protected to ensure a steady flow of water, to protect their biodiversity, and to acknowledge them as the mythical Mountains of the Moon.

In contrast, our approach, without challenging these values, set out to describe and validate a different set of values, the values the Banyarwenzururu had spent decades fighting for. To transform the lengthy investigation undertaken at Mburo into something simpler and quicker, a small team of park and project staff met with people from villages, trading centres and towns, government officials and community-based organizations. They asked these different groups, as simply as they could, what was important about the mountains to them, and what values they thought the national park should be helping to protect. Meetings were informal to encourage open discussion and – the organisers having learnt from similar participatory processes – were separate for men, women and young people. Experts in the local culture were also interviewed.

It was hard, however, for the local people to talk about the intangible values of their mountains. Most of the discussions started with details of how the mountains provided for their families. People talked about their farms, the soils, how the mountains brought the rain. They talked about their building poles and the firewood they collected, and about the nets, twines, baskets and utensils they made from the materials they gathered in the forest. They described the honey and mushrooms they collected, and the medicines they made from mountain plants. Some people talked about hunting for hyrax in the high moorland zone. Others described setting basket traps for fish in the streams or fishing in the deep pools.

When they had done, we asked if there were not other kinds of things that the mountains gave them – things not related to putting food on their tables or money in their pockets, but important nonetheless. Once the people understood what we were asking, it was as though we had removed a dam on one of their mountain streams, and a flood of information would rush out, bubbling and fizzing.

They told us of sites they visited to make remembrances, some of such significance that annual pilgrimages were made. There were sites on the edges of the villages as well as in the high mountains where offerings were made to gods and spirits, and rituals performed to cleanse the land and keep the people safe. We were told of spirits in the streams and in the forests; of places where people hid when tensions flared with lowlanders; of sites for coming-of-age ceremonies and circumcisions; and of the zones where people could go, and zones where no one should go.

They would go to the mountains to collect materials needed for blessings and ceremonies and to make offerings. Many household items, baskets and vessels and pots, had been replaced with market-bought goods, but there was nevertheless a desire, when engaging with the spirit world, to use traditional items made using the crafts of their ancestors. As the list grew, the sites, ceremonies and practices we were told about provided a window through which the sacred nature of the mountains could be seen.

Nyamuhanga, the Great Creator, had made the mountains and the forests and the people, but having created all things, would take little active interest in them or in the lives of people. Nyamuhanga made Nzururu, the spirit of snow, who had given his name to the people and the mountains. It is Kithasamba, though, the God of All Things, who engages with the people and takes interest in their concerns and their needs. Kithasamba lives in the snow and glaciers of the highest peaks, considered so sacred that they are rarely, if ever, visited. A host of lesser gods have their domains lower down the mountains, their presence creating different realms in which the people find their place – farming here, gathering medicines there, and hunting where and when these gods permit.

Respect for the gods remains part of the lives of the people and their use of the mountains and its resources. If they are to collect plants for weaving and building, or if they need to harvest herbs or cut bark or dig for roots and tubers with the power to heal, strict rules of behaviour apply. The tree or plant or place of collection is approached by climbing the most direct path, and the return must be by the same path. The harvester works facing up the

A shrine and offering made by Bakonjo hunters high in the mountains

mountain, standing below, almost in a position of supplication. They must make their harvest from that place, never circling, never moving around or expanding the location from which they harvest. People spending time in the forest are not free; they are not there to wander, to pick up this and cut that. Time spent in the forest is for studied, considered actions, approved by the spirits, and people are expected to demonstrate respect.

Any trip to the sacred zones of the mountains would be discussed and agreed with the ridge leader. He is the force of law in the community, both political and spiritual, his authority passed down to him through the king and the chiefs. The ridge leader, working with the guardians of the sacred sites, ensures the safety and prosperity of the people on his ridge. Overseeing that the proper process is followed when making use of the realms of the different gods is key to this. So too is the work of the site guardians, who undertake the rituals of cleansing and balancing. Hunters may climb to the highest valleys, just below the snowline, and trap and shoot the hyrax that live there, so long as the ridge leader approves and the correct offerings are made.

On my first visit to the high passes in 1981, during my first year in Uganda, I came across a shrine constructed in the form of a miniature thatched hut, with offerings of meat and herbs spiked on a pole next to it. I found the same shrines 30 years later, when I attended ceremonies performed to celebrate agreement over community access to a sacred site. On that day, as we climbed the steep trail from the village and entered the forest that sheltered the site, we were commanded to maintain a respectful quiet, to moderate our voices and avoid laughter. This was the way to behave in the forest under the eyes of the gods. the ridge leader blessed us on our departure and on our return. Too old to climb up with us, he gave us his permission to ascend the mountain and enter the forest, and his supervision of our visit remained essential.

Recognition by park management of the role played by the ridge leaders was perhaps the most exciting development of the project as a whole, demonstrating what a cultural approach could achieve. The authority that came down to the ridge leaders from Kithasamba was accepted as part of managing access to the park and its resources. The rangers had limited appetite for controlling the borders of the park and were ineffective at managing access to forest resources under the agreements signed a decade before. This was partly because of the remote locations and rugged terrain, and partly because villagers held the rangers in low regard compared to the ridge leaders.

When Uganda became a republic in 1966, the traditional authority and functions of the ridge leaders was removed from them, and when in 1991the park was declared they stepped back even further. The result was that from then on, nobody managed access to the forest. But for the remoteness of those areas their resources, once so carefully regulated by the ridge leaders, would have collapsed.

The park managers recognised the lack of control as a problem, as it undermined their authority. The ridge leaders saw it as a problem for different reasons; they might worry about the overuse of some resources, but of far greater concern was the lack of respect to the gods generated by the lack of control. They could not reverse climate change, which

A sacred site where cleansing ceremonies are performed for the ridge by its guardians

they feared would melt the glacier home of Kithasamba,[177] but they could make sure that those accessing the forest and the sacred sites behaved properly. Though not intending to deliver conservation as understood by the authorities, from the 2010s the ridge leaders once again began to promote their role in sustaining and strengthening the natural and social order.

There was a rapid evolution of the engagement between the park and the people. A formal agreement was signed with the king representing the Rwenzururu nation,[178] under which the authorities recognised the sacredness of the mountains, agreed they were intrinsic to the park, and accepted that they should take practical steps to give meaning to them in the park's management. The local importance of the burial caves of the kings was accepted, and annual pilgrimage to the site facilitated. The site itself was developed to accommodate large numbers of pilgrims, and a museum was established to add protection to the site. Other sacred sites inside the park were also recognised, and access to them negotiated.

Perhaps the most exciting development for me was that rangers and ridge leaders began working together to manage access to the forest and its cultural and natural resources. The park managers accepted that respect for the ridge leaders was stronger on the ground than for the rangers, especially after the government had formally recognized the king.

The directors of the Uganda Wildlife Authority were excited by these achievements in the Rwenzoris. Could the cultural approach help resolve problems in western Uganda's Western Rift? they asked. As mentioned in Chapter 8, when the forest parks of Bwindi, Mgahinga and Semliki had been established in the early 1990s the treatment meted out to the Batwa hunter-gatherer people was appallingly damaging. Since then, despite many interventions and initiatives designed to support them, little has actually been achieved; the Batwa were and continue to be amongst the least educated and the poorest, unhealthiest and most destitute people in Uganda. The wildlife authority did not formally accept responsibility for their desperate plight, but it did want to improve things.

---

177 Tourists, accompanied by local guides and porters, now regularly encroach on the domain of Kithasamba. Quite a few of the Mountain People actually blame climate change and its increasingly disquieting implications for their gods and their lives – including important permanent rivers fed by meltwater from the glaciers running dry for parts of the year – on the regular presence of the climbers on the highest peaks.

178 President Museveni began restoring the kingships that had been disbanded in 1966, when Uganda was declared a republic by the first president, Milton Obote. The first to be allowed to return and establish his royal court as a cultural institution was the Kabaka of Buganda; his inauguration and the accompanying celebrations made the international news. The kingships of Toro, Busoga and Bunyoro were restored and finally, the Omusinga of the Rwenzururu Kingdom was restored in 2008. After a number of years of calm under the Omusinga, tensions grew with other ethnic groups and the government until armed conflict erupted in 2017 over rumours that the Omusinga was once again planning to secede from Uganda.

Above: Descending through storms into the Albertine Rift and Semliki Forest

Left: Batwa, famed for their music and dance, are marginalized and live in poverty (copyright Henry Busulwa/FFI)

As with other peoples, the Batwa's way of life had evolved over time, responding to changes that provided challenges and opportunities. Batwa were the earliest peoples to inhabit the great forests of the Western rift, but under the axes of the farming peoples who had arrived in the region much later, some 3,000 years ago, the forests shrank. By the start of the 20th century, when British administrators first arrived in western Uganda, relatively little forest remained there. In neighbouring Rwanda, too, the forest had mainly vanished by then, and the Batwa there had become makers of pots, respected as a distinct ethnic group within Rwandan society.

In Uganda, however, the forest was still connected to the vast forest that stretched across the Congo basin to the west. There the Batwa lived freely, hunting and gathering as they liked, engaging on their own terms with the farming communities that nibbled away at the forest edge. They traded valuable products of the forests, especially meat, that their superlative skills allowed them to acquire, for iron pots, steel knives, axes, cloth and jewellery. The Batwa would take work as labourers or livestock herders to get the goods they wanted – but they would return to the forest again when it called them.

Initially little changed with the arrival of the British. The Batwa continued to make their extended perambulations through the forest, collecting foods and herbs and fibres for subsistence and trade. It was not long, however, before the British and the settled farming peoples, who saw the Batwa as poor and backward, determined to bring them out of their primitive condition by showing them the benefits that went with a settled life.

The Batwa, however, were happy as they were. They described the forest as their father, providing for all their needs. They pitied their unfortunate neighbours, forced to live outside the gentle forest, and to labour in the full heat of the sun to meet their needs. But when Semliki Forest Reserve was established in 1932 and Bwindi Forest Reserve in 1942, the Batwa were ordered to cease their roaming in the forests and were obliged to settle in villages at the forest edge.

The change that occurred when the forest reserves became national parks – from the point of view of the parks authority at any rate – was merely that the Batwa villages were moved from just inside the forests to just outside them. The distance was short, and the move was simply completing a settlement process that had begun long before. The difference for the Batwa, however, was not where they lived but *how* they lived; with the creation of the national parks they were no longer allowed to make their living from the forest. It is true that it was not the wildlife authority alone that was responsible for what had befallen the Batwa, or that it was only the Batwa that suffered loss when the parks were declared – but what was done to them was especially terrible. Though the Uganda Wildlife Authority did not officially acknowledge this, the need to address the situation had become clear by the time they asked for our help, 20 years after the parks were established.

Though incremental change had been occurring for centuries, until the parks had been created the changes had remained largely in the control of the Batwa themselves, so the forest continued to define them both as individuals and as a people.

Then the British had forced a more rapid and disempowering process of change. And it was the creation of the national parks that changed everything, once and for all. The lives of the Batwa were suddenly and irreversibly transformed. Now they were prohibited from hunting and gathering. Now they were not tolerated in the forest at all. They were prevented from visiting the graves of their ancestors, and they were prohibited from dancing and singing at their sacred sites. Thrust into the far from tender embrace of the farming communities that despised them, they could find no refuge from the people that had become their often cruel masters.[179]

---

179 The forest peoples, historically called pygmies because of their small stature, have been subject to mistreatment by dominant Bantu tribes in central Africa, including forced removals, slavery and genocide. https://en.wikipedia.org/wiki/Pygmy_peoples

Their source of livelihood denied to them, they were forced to live as farmers, but without any lands of their own and without the knowledge they needed they were exposed to the most severe exploitation. In addition, they suffered from the ills of all people forced to abandon their gods and their way of life by a sudden and irresistible shock of imposed change. They fell prey to despair and lack of hope, and also to the scourges of alcohol and domestic violence that so often follow when social and cultural institutions fail.

The Batwa saw no future for themselves and they began to lose the positive sense of what made them a people. While recognising the uniqueness of oneself and one's people can be powerful and define a sense of one's worth, the labelling of people as 'other' can be terrible when used to justify prejudice and exclusion. When the Batwa began to lose the sense of themselves – different and other from their farming neighbours – they began to lose the sense of their place in the universe.

The national government, having established the new parks in the 1990s, failed in its responsibilities towards the Batwa, while local government, representing the interests of the farming communities, largely ignored their plight. Conservation organizations were conspicuous by their absence. Well-meaning Christian churches, stepping forward to help, were joined by development charities, but despite the good intentions little was achieved.

Some of the worst privations were lifted. But although some Batwa children were supported to attend school, some families were given access to land, and some were taught to farm, by and large success was limited, and the sad truth is that Uganda's Batwa community shrank year by year, while their specialized knowledge of the forest and their cultural identity eroded steadily away.

The reasons underlying the persistent failures of the initiatives to improve Batwa lives are complex. The overriding problem, however, seems to be that most projects were grounded in the assumption that Batwa should stop being Batwa. The Batwa needed to stop thinking about the forest, stop yearning for the forest, and especially stop going into the forest. They must give up hunting and gathering to become farmers; give up their forest gods to become Christians; give up their knowledge of the forest in exchange for the ways of their neighbours. The help offered made no effort to respond to the Batwa's understanding of themselves, and although the Batwa did not actively resist the help offered, that help, unsurprisingly, simply failed to germinate.

The Batwa do not, by and large, demand to return to the forest or to live as they had done 1,000 years, or even just 50 years, before. The Batwa can see the value of education and health services and piped water as well as anyone. But they do not wish to give up their special relationship with the forest and its animals and plants, and they do not see

*The Impenetrable Forest lying in the western Albertine Rift Valley
is part of the ancestral home of the Batwa people*

why they should. It is their connection to the forest that defines them and which, in a way, defines the forest too.

It was this interconnection between the Batwa and the forest that our cultural values approach focused on. With funding this time from Britain's Darwin Initiative, and working closely together once again, Arthur and I proposed to express, represent and engage this ancient entanglement between people and nature as a valuable benefit for the Batwa and the parks. The project would support steps, in ways both practical and symbolic, to reverse the exclusion of the Batwa.

We promoted the employment of Batwa not as labourers or porters, but for their unrivalled knowledge of the forest, their capacity to read the forest, their ability to navigate its moods and secrets.[180] The Batwa would thus strengthen any research programme, whether it was assessing biodiversity, mapping ecosystems, or understanding the movements and seasons of the animals. The Batwa would make the best guides and interpreters of the forest for tourists, whether tracking gorillas, searching out rare forest birds, or simply enjoying the complexity and beauty of the ancient forests. Only Batwa could give insights into their own relationships with the forest, how they lived in it and from it, and how the forest made them the people they are.

Helping Batwa to find jobs like these would not only contribute to reducing their material poverty but also provide reasons for retaining their specialist knowledge and

---

180 The single mutwa employed by the Uganda Wildlife Authority worked in Bwindi. He was fondly nicknamed Geepee-ess (GPS – Geographic Positioning System) by his colleagues because of his unerring ability to know exactly where in the impenetrable forest he was at all times.

A Batwa guide piloting the development of the Batwa Forest Experience (copyright Pamela Nabukenya/FFI)

skills. Such jobs would also give their special knowledge a new validity. Equally important, this approach would demonstrate the capacity of the Batwa to help achieve the conservation objectives of the parks – objectives that the Batwa were passionate about themselves, and which they feared the park authorities were failing to deliver.

Of central importance to the project was the issue of access to the forest. Access could not be understood as simply allowing Batwa into the forest to work, however valuable this might be. Access entailed a recognition of the need for the Batwa to be in the forest in order to continue being Batwa. Reaching a proper understanding of access would require the authorities to accept a different perspective not only of the forest they protected but also of the place of the Batwa in it.

The forest was more important to the Batwa than to anyone, and if they were allowed to, they would become its greatest supporters and protectors. Certainly, there would need to be compromises to line up the interests of the parties, but once achieved the benefits would be enormous all round. The Batwa would once again be people of the forest, dancing and singing their prayers and gratitude to it as their ancestors had done. And the forest would be richer for being imbued with their unique place in it, filled once again with the spirit of the people of the forest.

The project delivered valuable outputs. Batwa sat with the park authorities to discuss decisions about the project and the parks. The Uganda Wildlife Authority and our donors were delighted. The Batwa, too, were happy with the support we provided to improve incomes from farming and handicrafts and strengthen their voice in negotiations with the authorities.

The main outcome from the project was a trail along which Batwa could guide tourists, giving them an experience of the forest as they, the Batwa, experienced it. The path through the forest was prepared and Batwa were trained to work with tourists. Batwa guides were issued uniforms distinct from those of the park rangers, and bearing insignia that highlighted their unique skills and knowledge.[181]

---

181 The badges showed a squirrel and a swallow held in cupped hands. These motifs were chosen by the Batwa to represent the skills required before a mutwa could marry. He had to present to the father of his intended bride a live squirrel and a live swallow, which he had caught in his hands. They would then be released back to the forest as part of the acceptance ceremony.

Thus the Batwa Forest Experience was commissioned – but it did not thrive. The park authorities, apparently, could not allow Batwa to lead the enterprise, even with project support. Too many demands were made to share the revenues. And finally it was the accompanying rangers and translators that killed the project. They could not accept that it was the Batwa who had the knowledge that provided the real interest for the tourists, nor that it was the Batwa values that lay at the heart of the enterprise.

The project did achieve improvements for Batwa, but nobody in officialdom responded to the cultural approach of the project: the notion of making the forest complete by reflecting Batwa values; giving

A Batwa guide demonstrating her forest knowledge

validity to the values, practices and knowledge of the Batwa; and allowing Batwa to be Batwa. The park authorities could apparently only see them as another group of people demanding access to the parks. Most damaging for the project was the rejection of the need to treat Batwa as exceptions to the park authorities' many rules. On the critical requirement of access to the forest, for example, this inflexibility meant that Batwa could only enter the forest accompanied and for park reasons, but never alone and for their own reasons. Our assertions that the parks should search for solutions to allow the Batwa to regain their place in the forest, perform their ancient ceremonies and pray to their gods fell on deaf ears.

9

Nature will save us

I spent almost four decades helping make protected areas work better, in Africa, in Asia and now back home in England, on Ashdown Forest. I remain convinced that protected areas are essential to save at least some of our planet's natural wonders and even to save us and our descendants.

I cannot, however, reconcile myself to the disturbing downsides of those protected areas, and even now, in 2023, I feel less and less sure about the best course to follow. Exposure to the complexity and difficulties of conservation over my career has eroded the certainty of youth, but perhaps it is the nature of the endeavour itself that disturbs – not only the dispiriting intimations of failure, but the internal contradictions associated with protected areas.

I have observed the exceptional commitment of individuals, the considerable financial investments of organizations, and the personal sacrifices made for them. I am aware of the loss of life of the frontline wardens and rangers working to safeguard wild animals from criminal gangs, the deaths of villagers caught in the crossfire, and the many costs imposed in their name.

But despite this, the gradual yet steady erosion and degradation of parks and the natural world continues, and at the time of writing seems set to accelerate.

Conservationists have responded differently to these difficulties. Some want to see greater emphasis on laws and regulations, a return to stricter models of conservation. They would pull back from community approaches. Stronger punitive policing has powerful voices of support.[182] Returning to an earlier and apparently simpler time would seem to put aside the moral and practical dilemmas of protected areas. As before, the importance of protecting nature would trump other considerations. Proponents of a return to 'fortress conservation' argue that working with communities has failed to turn the tide and that

---

182 In 2014, a conference convened by the British government, attended by world leaders and addressed by Prince Charles, received global media cover. Calls were made to intensify action against criminal poaching gangs, and ivory and rhino horn traders.

the land carved out for nature would be lost in any attempt to satisfy rising and widening demands on them. The collateral damage of the protected areas – whether from the anti-poaching campaigns that net the small-time local users, reductions in resource use to sustainable levels, or reserving land exclusively for biodiversity – seems always to be visited on those with the least power and the least wealth.

Most conservationists, however, especially those working on the front line, continue to emphasize the need to work with, rather than against, communities. If I agree with any of the current proposals, I agree with this. But I believe we must find better ways to achieve it than we have to date. Ensuring that parks meet the needs and aspirations of their neighbours as far as possible remains the key for many. Others wish to make neighbours into partners in the fullest sense.

It often seems, though, that the advocates of these approaches engage with them as objectives for conservation to achieve, rather than as strategies for achieving conservation. Pursuing them as objectives seems to recast conservation as part of the sustainable development agenda. This is increasingly evident, in the demands of the government agencies, the grant-making foundations and even the private donors, that conservation must produce economic development outcomes too. This is problematic. Conservation delivers many benefits – both material and intangible – but conservation cannot always deliver the economic benefits pursued by development aid agendas.

These days I may be certain about little regarding conservation, but I know that the world will be a sadder, less inspiring home for humanity without its elephants, silver-studded blue butterflies, tokay geckos, giant redwoods, and all creatures, plants, fungi and algae, great and small.

And I believe that the children of the future will condemn us if we consign to them a depleted and less eloquent world. To avoid this, we must find a different way to think about nature and find better ways to live on this planet. This means considering again, and more deeply, what nature itself might be, and what it might be for. This sounds obscure and academic, a recipe for introspection and hazy discussions. However, new kinds of expert are needed to lead the thinking of ways to protect nature through the 21st century.

Conservation policy and practice, and the political and social context in which it is pursued today, is different from what it was in the early 1980s, in the heady days I spent cruising Uganda's national parks, or in the 1990s, learning about the Enyemibwa in Mburo, or in the 2010s, failing to engage colleagues with the cultural needs of the Batwa communities.

But despite its achievements, conservation is not working. We all know about the losses of our planet's multifarious forms of life under the heavy hand of humanity.

Everywhere nature is in retreat. The prodigious herds and shoals and flights of animals that our recent ancestors took for granted are for the most part gone. Vistas of forests, grasslands, wetlands, mountain ranges stretching unbroken to the horizon, are now carved by highways and railways, fields and towns. Factories and vehicles still darken our skies with oily clouds pumped from chimneys and exhaust pipes. Our soils are eroded, our waters polluted, and the oceans full of floating – and outweighing them ten times over, sunken, hence invisible to us – plastics.

Bemoaning these facts and denouncing those responsible – most of us, in truth – is not helpful. Separating the unavoidable impacts of human existence on our planet from wanton and unthinking damage is not easy. Sustaining human life is not merely about survival. When talking about life, we are not talking about biology alone but culture too: life, full, diverse and rewarding, as opposed to survival, limited, constrained and exhausting. There is a world of difference between them, but what precisely makes the difference is uncertain. It is relatively easy to agree how nature contributes to our survival, but agreeing where and how it contributes to lives requires a broader analysis.

Demands for social and economic development put arguments for conservation on slippery ground. Development may be about improving lives, but the target is subjective. Access to electricity is transformative – but should all demand Nike trainers? Or long-horned Ankole cattle? We can fall back on 'meeting people's needs', as Arthur, Moses and I tried to do in Mburo with our SCIP projects but needs are not easy to define or to separate from wants. And if we can indeed separate them, is it reasonable to expect or even possible that all desires can be met?

The fact that humanity ultimately depends on nature suggests that the decision to protect it, or at least to use it sustainably, should be an easy one to take. I am hopeful that humanity will act to ensure that nature can go on sustaining lives, and hopefully good lives. But nature is an almost infinite number of interconnected things, in terms of the living organisms and ecosystems themselves and in how we conceive them and use them.

Agreeing to protect nature may not mean protecting all nature. That some elements of nature inflict harm on us suggests the opposite, as the Covid 19 pandemic of 2020–22 made abundantly clear. If we cannot or do not want to save everything, differences must arise over what to save and what can be let go. The same goes for the kinds of use found acceptable to subject nature to. Do all agree to keep elephants in our world? And in our own back yard? And for what? Maintaining ecological processes? Photographic safaris? Leather wallets?

If you have ever spent time in a warm climate, you will surely have encountered geckos, small lizards with magically grabby feet, bobbing their heads and chirping. As a boy I was enthralled by Gerald Durrell's descriptions of geckos scudding across the walls of his bedroom in Corfu, stalking moths and mosquitos – and on one celebrated and bloody night, a praying mantis. During my years in Africa and Asia, geckos have always delighted me.

In Mburo in the 1990s, tropical house geckos lived in my tent, the flapping canvas walls creating no difficulty for their sticky feet. When I lived for six months on the Jiddat al-Harasis, the stony desert of central Oman, yellow-bellied house geckos had taken a ride on the prefabricated house I lived in, and we happily shared the veranda as the sun sank and the day cooled. Geckos covered the walls of our house in Hanoi, and on most days during my time in Bali a giant Tokay gecko sedately passed above the window where I sat working. All exquisite creatures, they kept me company, but constructing an honest argument for how they did, or even might, improve the lives of the poor farmers I worked with was, and remains, beyond me. I might speculate on medical properties as yet undiscovered, or on inventions based on their wall-walking footpads,[183] or their value in the pet trade, or the undoubted fact of their eating mosquitos and other insects in the house. None of these would be likely to help much, however, in hard economic or social terms. In any case, for me none of these arguments would be more compelling than their inimitable quirkiness and the poetry of their clicking cries and robotic dances.

These are real values, but they don't balance economic equations. Without other arguments to save them, geckos will probably not make it into the ark. And if they don't make it aboard, the countless other animals and plants that enrich our lives but have no obvious or immediate economic value will be lost too.

What about animals like lions and tigers that people may feel a stronger connection to than geckos? I have admired the sensuous strength of tigers since watching one prowl a snowy enclosure at Whipsnade Zoo as a boy. Much as I admire them, and much as I know that tigers are on the edge of extinction, it would be a stretch for me to assert, and difficult to prove, that they are necessary to human survival. And of course people living where tigers still roam, taking a goat here, the occasional unwary person there, might make a compelling argument to the contrary.

Should I brush aside their reservations, arguing that a creature so fine must be protected, or should I cobble together an argument around tourism and jobs for poor communities, or the magical properties of their bones? And then, what of all the millions of unremarked and apparently unremarkable creatures, if any creature can truly be described as unremarkable? There is no escaping the hard choices to be made, but agreeing the basis for these choices will be difficult. Science and economics can guide us, but I hope that the many values of nature, informed by cultures with all their relativity, complexity and messiness, will be central to the process.

The simple idea of working with communities, rather than being in constant opposition to them, has done much to change conservation for the better, but the approach has been limited by resting so squarely on economic and science-based approaches. The vexed

---

183 There really is such an invention, though the question of how this helps communities living around protected areas remains. http://www.sciencemag.org/news/2014/11/gecko-inspired-adhesives-allow-people-climb-walls

question of how parks that exclude their neighbours can become parks that engage them has come to mean 'how much can a park provide?' How many jobs, how many kilos of bush meat, how much development, how much economic growth? But nature provides other things too, and creating access to these can create reasons for people to support conservation, just as denying access to them can cause hardship and loss and generate conflict. It is time to explore a broader range of resources that parks can deliver to their neighbours, cultural as well as material.

In 1988, while worked on a social survey of hunting in the newly established Korup National Park in Cameroon, I walked through the market in Mundemba Town, on the edge of the park. The experience was not unlike browsing the food hall of a department store, though a lot noisier, more colourful and friendlier. The variety of goods offered for sale was seductive, the cacophony of calling and haggling, the banging and chopping and pounding, was disorienting, and the aromas were sometimes disturbing, sometimes transporting.

I marvelled at the mounds of leafy vegetables, looking greener and more delicious than I thought leaves could ever look, some finely chopped as though run through a paper shredder, some stacked like leathery plates, others piled in lush, fragrant pyramids. There were enamel basins full of churning masses of polished black-red termites. Nuts, berries and fruits were arranged in handfuls or in stacks, or presented in baskets to be measured out by the cup. There were murky bottles of oils – gold, yellow, red – pressed from seeds and nuts collected from forest clearings. Smoked catfish lay side by side with dried bush meat and collections of glistening discs that were desiccated giant land snails. Sometimes whole animals could be found with singed, sooty skins; cane rat, blue duiker, red-tailed monkey.

All this bewildering variety of foods had been collected in the forest. The market sold garden fruits and vegetables too, as well as pig and goat meat, but the forest foods were special. The favoured dishes were based on them. Powdered bush mango and forest oils gave sauces a glutinous quality that was much sought after.[184] The defining dish, pepe soup, was not only expected to include a knob of smoky bush meat – farmed meat was a poor substitute – but also numerous other ingredients such as herbs and leaves and oils from the forest, needed for perfection. Increasingly these ingredients were only to be found in Korup National Park, but the tacit acceptance that had allowed their collection was giving way to an insistence on adhering to the laws that made this illegal.

The multiple commodities on sale in the market, as well as their gathering, processing and transport, had economic importance, but their true value went far beyond, and was

---

184 Perhaps the most well-known version, familiar to all along the West African coast, is okra soup. Admittedly in this case made from a cultivated vegetable, the dish is made by boiling okra, producing a dish that to the western palette can be best described as slippery.

entwined with what it meant to be Cameroonian. Highlighting the link between Korup National Park and the ingredients essential for pepe soup – the national dish – might have stimulated more interest amongst local communities than tales of Korup's financial value. But this would have required the park managers to be open to the idea of communities harvesting forest foods – something that they could not agree to. Just as in Mburo, where the wardens could not accept that cattle could be part of a national park, the wardens of Korup could not accept that hunting and gathering could be part of their new park. But in both cases, integrating these powerful cultural links between people and place would have generated a local resonance and a connection that the sterile scientific and economic arguments for the parks could not.

In South Africa, when looking at the role of wildlife reserves in the local economy, I asked Zulu communities what they thought about the reserves they lived around. Men told me that they were important because of the animals they could hunt. I pointed out that they were not allowed to hunt, but this seemed to make no difference to their position. It was not specifically that they wanted meat. They did, but they wanted to hunt more, and in particular they wanted to hunt with their chiefs.

Before the land had been taken to become nature reserves 100 years earlier, hunting had been controlled by the chiefs. As a mark of respect, warriors would be called to hunt together. And after hunting they would feast. It was an act of recognition that built connections, binding the chiefs to their warriors and the people to their place. The nutritional value of the meat would have been incidental; it was the act of hunting and feasting with age-mates that was important. It was the desire to experience this cultural history, not the meat, that kept alive the wish to hunt. It was this that gave the reserves value in their minds, even though their wishes were denied, just as Mburo was intrinsically valuable to Bahima as a connection to their ancestors and their Enyemibwa.

Culture and practices imbue nature and place with value. To the Bahima, Mburo in the absence of their Ankole cattle was as nothing. Values may be dissimilar or contradictory, but will be reflected from experiences of the place. For many, protected areas are places of plants and wildlife, and are locations of beauty, tranquillity and adventure. Special features and species, especially if they are large, charismatic or rare, attract particular attention. To some, obligations under international conventions to protect 'national heritage' will be important.[185] Most parks were not originally created to conserve biodiversity, but the concept has become so important that it now defines their purpose.[186] Ecologists, meanwhile, talk about the functions of ecosystems and the goods and services provided

---

185 Conservationists often describe countries that harbour populations of a certain species as a 'range state', and where the species is an iconic one there can be real national shame on being removed from the list. Cambodia, for example, has accepted that it has lost its tigers, but has committee d itself to bringing them back. Neighbouring Vietnam has, however, never accepted that it has lost its tigers.

186 A number of global assessments of biodiversity have been undertaken to define ways to identify the most important sites for conservation. These include biodiversity hotspots, eco regions and key biodiversity areas, all similar in their intent though different in their science and definitions.

by the parks.[187] Each of us will look at a park through the lens of our values and interests, and our attitude towards it will depend greatly on whether or not we see these reflected. It may not be possible to agree what parks are for, the answer dependent on the perspectives of who answers – but there will be no shortage of interest in the question.

The values of a protected area, then, are specific to the user. What is an essential attribute for one party may be irrelevant to another. As part of my research at Mburo I looked at comments made in the visitor books left at the park gates. They expressed very different experiences, revealing widely divergent expectations of their writers' visits. Stripping out the common excitement at the wildlife, there were surprisingly clear differences between nationalities. Americans often gave thanks to God for the glory and wonder of his creations. British visitors, strangely, seemed mainly interested in the park's facilities – the state of the tracks and whether the showers were warm. Continental Europeans would write of the views and vistas of the park. The comments of Ugandans tended to reflect on the park's achievements or lack of them.

By comparison, Mburo's significance to the Bahima, defining as it does their identity, made their relationship with the place more resonant than that of any other group. It would be reasonable to expect this to make their interests pre-eminent in the park's design and management – but not so. It was others who, with greater power and authority, controlled these processes, and they were reluctant to share them. It is conservationists, who see parks as score sheets for species (rare species score double, endemics, triple) and economists, counting the profits and losses, who determine the expression of worth. Despite what seems a rather narrow set of interests, these are the people that dominate the descriptions and justifications of parks, including those of Mburo.

If I found it difficult to express my feelings for Mburo, or for nature generally, how were the Bahima to convey their experiences to me? I could expect little more than a hazy impression of their internal world, and vice versa. But this level of identification with another's emotional response to nature should be enough to accept and respect its validity. Indeed, there are good reasons for not going beyond this to avoid the appearance of adopting or claiming parts of another's culture.

This concern over cultural appropriation cannot be ignored so long as unequal power affects the relations between peoples. Although borrowings and exchanges

187 The Ecosystem Services framework, developed to provide economic justifications for conserving nature based on the value of the goods and services it provides to humanity, generally describes four kinds of service: provisioning services such as the fish, fowl and fibres we consume; regulating services such as flood controls; cultural services, a catchall for all the values of nature that are intangible and difficult to measure but that cannot be ignored; and supporting services, the basic functioning of ecosystems that underpins the delivery of all the others.

between cultures have always occurred, allowing those cultures to evolve and adapt, the appropriation of beliefs and practices by conservation interests threatens to weaken them and undermine their potential to support the practice of conservation. The meanings and functions of the traditional controls over the use of particular places and resources need to be respected for what they are rather than co-opted to serve conservation. Whether taboos, totems or royal privilege, their purpose lies within the cultures that established them and they should not be twisted to meet the interests of modern conservation.

Such mechanisms, which protect locations, animals and plants, can deliver multiple outcomes. They might ensure that the catchment of a river, stream or spring is not cultivated, thus maximising the likelihood of a constant supply of water; they might ensure that alternative grazing is available during droughts, or ensure a supply of wild meat during the hungry months of the farming cycle. These outcomes might be the intention of limiting access to a place or a resource, or the incidental byproducts of something entirely different. The sustained supply of resources might go quite unrecognized as an outcome of controls imposed for other reasons.

Taboos against harming clan totems – mostly wild animals – in Uganda, for example, may have helped protect those animals. The primary intention of those taboos, however, was to reinforce clan identity, ensuring that marriages were outside the clan and therefore outside the immediate social group. This helped create extensive social and economic networks and avoided interbreeding within small communities.

Nshara was set aside as grazing for the royal herds. These magnificent herds demonstrated Ankole's excellence in cattle-keeping, strengthening morale and helping the tribe compete with the neighbouring tribes with similar ambitions. The control that Zulu chiefs exerted over hunting and the hunting grounds helped weld warriors into a loyal and cohesive military force. None of these outcomes are remotely related to conserving resources, even though through them conservation was undeniably achieved.

It is tempting to describe these traditions and practices as indigenous conservation and to label the people as instinctive conservationists. It is common to hear tribal peoples described as the natural guardians of nature, and to hear their practices and beliefs interpreted in the language of conservation. But many of the practices that conservationists happily describe as protecting a species or a place are in all probability undertaken by the indigenous peoples in order to achieve something else, such as displaying reverence to a god or respect for the ancestors, or channelling planetary energies, or focusing the powers of the spirit world.

Such sacred or spiritual outcomes are likely to impose some limits on day-to-day life. It is easy for conservation to appropriate sacred practices and fit them into our worldview. But encouraging them in the name of conservation disempowers communities and their institutions. Our insistence that traditional societies protect nature just as we do, and that they should work towards the same outcomes as our protected areas, ultimately undermines their beliefs.[188] This is not to say that the local people never set aside areas with the clear intention of reserving resources for hard years or protecting water sources. There are plenty of examples of these– but traditional practices that happen to result in conservation should not be confused with practices intended to achieve conservation *per se*. Intention is key.

During the Cultural Values project I learned of a community just south of Mburo who protected pools on the Akagera River. Keen to give the practices more importance the team began to describe them and the beliefs associated with the pools as mechanisms for protecting fish-spawning sites.[189] The pools certainly were important for sustaining the fishing, but the community protected them and hedged access to them with rituals and ceremonies because they were sacred. Ceremonies were performed at the pools to bring rain, ensure good harvests and catches, and assure good health. The distinction is important. The pools were sites for ceremonies that protected the *people*, not the fish.

Most of the world's cultures are monistic. They see the cosmos as comprising an indivisible whole, a single entity, and they understand humanity and nature as inseparable, parts of the same thing. They recognize no divide between the natural and unnatural, between the human and non-human. This being the case, many of the world's languages do not even have a word for 'nature' because it is not separate from humanity.

This is different from the idea that lies at the core of western culture, which understands the universe as divided into opposing principles; mind and matter, good and evil, natural and unnatural. Dualism is central to science and technology, and therefore to the world of conservation too. It is so fundamental to western culture that it is easy to forget that the west, in seeing existence in this way, is very much in the minority.

---

188 Paige West discusses this understanding in *Conservation Is Our Government Now*. Based on an eight-year study of a conservation initiative and a local community she demonstrates how a failure to engage with the worldview of the people, the simplification and misrepresentation of their beliefs, and an insistence on interpreting their relationship with nature in western scientific and economic terms led to the failure of the conservation objectives of the project.
Conservation Is Our Government Now: The Politics of Ecology in Papua New Guinea. West, P. (2006) Duke University Press.

189 A conservation project presented information collected about a unique system of ownership and management of fishing pools, both natural and constructed, on rivers flowing into Lake Victoria. Though most of the pools were owned and handed down through families and had been for centuries, some pools were set aside for the performance of ceremonies, especially rainmaking ceremonies. Without good rains the biannual migration of fish would not occur on which the fishing depended. The conservation experts noted that communities carried out the ceremonies to protect the pools for sustainable fish production, which community members confirmed. But this explanation represents a distortion of the actual purposes of the ceremonies. The pools were set aside from day-to-day fishing to allow for rituals and ceremonies carried out for the purposes of making rain, respecting the ancestors, placating spirits and so on. The conservationists were keen to translate the beliefs and practices of the communities into their own conceptual framework, while the communities had learned to agree with the experts in the expectation of financial and political support.

Humanity understands its place in the physical and biological world in uncountable ways. These ways of seeing place us in the land that we transform into landscapes through our beliefs. Thus, while the Beautiful Land is understood by the Bahima to be the work of the Bachwezi, their god-like ancestors who had carved the valleys, mounded the hills and filled the swamps, Jews, Christians and Muslims live in a world created by God in seven days, as described in their holy books, and aboriginal Australians conceptualize their homelands as the dreamings of their ancestors;

These are tales that people have told over the millennia to make sense of their worlds, and for most people they remain as real and full of meaning as the stories told by science. Whether understood as myths, metaphors or literal truth, these explanations of origin, place and connection provide the essential meanings of life for the majority of the people of the world, and for them are truths that explain what the world is, where it came from and how it works.

But many people, scientists in particular, do not believe these stories. Science explains the world to conservationists, telling us how the natural world came about and describing what it is. History and power as well as practical utility have determined that it is mostly these understandings that are reflected in the parks and reserves designed by conservationists. This is why most of them believe their decisions relating to those parks are rational, being entirely based on science. Suggest otherwise – introducing, perhaps, the idea that they may be motivated by other, emotive, values – and you are likely to be met with a puzzled but firm denial.

In fairness, conservationists have always had to make arguments for protecting a species or a place in terms that governments are likely to respond to, and informed decision-making means couching arguments within the twin worldviews of science and economics. It is hardly surprising, then that we forget other realities exist and see science as the truth, rather than just one way of seeing the world. As a result, conservation tends to see itself as instructing, explaining and building arguments for saving nature on the reasons proposed by science.

The lack of interest in and engagement with the ideas of modern conservation about nature and its protection that I encountered when working with communities suggested an urgent need for a different message expressed in different language. I had to learn to talk more openly, honestly and inclusively about the purposes of conservation, and avoid suggesting there was a single, simple answer. Indeed, accepting multiple answers based on diverse values and perspectives began to seem a more reasonable proposition, more truthful, and more likely to secure support.

I do not have a singular relationship with nature. I recognize its values in the contributions it makes to human wellbeing, putting food in our mouths and a roof above our heads. I believe that nature inspires, creates a sense of wonder, and seeds minds with visions and revelations. I see that nature engenders a sense of being, belonging, and connection to place and self. I also hold that despite these connections for humanity, nature exists outside the human realm too, existing solely for itself.

I have not constructed the belief that nature is all these things in order to bring together an increasingly fractured and fractious conservation movement, or to co-opt the perspectives of others. It simply helps me point out that conservation can be achieved more effectively, more reliably and more equitably by being broader, more open and more inclusive. I cannot easily set aside the dualism that lies at the heart of my culture, but fortunately I don't need to do that. I just need to place my own beliefs no higher and no lower than the myriad ways in which others experience their universe.

To me, the arguments for integrating cultural values into the design and management of protected areas seem neither controversial nor difficult. I don't suggest that they represent the only answer to the problems of conservation, but I have no doubt that they would make an important contribution. But the idea of placing culture at the centre of initiatives to build support for conservation is viewed with suspicion by the planners and policy makers, who demand proofs of its efficacy. In my opinion, it is of course fair to challenge new ideas – but it is equally fair to challenge the status quo and ask whether the model based on science and economics is working.

The facts suggest that it is not. This being so, the situation being so urgent, and the significance of human connections to nature so evident, are such proofs really necessary? The clarity of the logic seems too compelling to be ignored. Conservation planners and policy makers just need to get on with it. But they don't.

Why is there apparently so little interest in culture within the mainstream conservation organizations?[190]

In academia there is plenty of interest; there it is something of a hot topic. Then there is the biocultural conservation movement, which holds that culture and nature are inextricably linked and that all conservation initiatives should pursue both objectives in tandem. Many commentators talk of an imperative to recover humanity's ancient relationships with nature to save ourselves from ruin. Indigenous Peoples movements, too, highlight the links between culture and conservation, arguing on one hand that conservation has ridden roughshod

---

190 While writing my thesis on the cultural nature of the conflict between the Bahima and the park managers I observed at Lake Mburo, I wrote and submitted a paper to a top conservation journal suggesting that mainstream conservation was mistaken in its exclusive focus on science and economics, and that it was overlooking the importance of culture as a mechanism for building support for protected areas. The three anonymous reviewers were unanimous in condemning the idea and the paper, though one grudgingly noted that I was right to express my views, however misguided. I was invited to submit a much-reduced paper, which was published as a note under the title, 'Cultural values: A forgotten strategy for building community support for protected areas in Africa' in 2001. Though there is broad acceptance of links between culture and conservation today, mainstream conservation has yet to absorb it at a practical level, which has always been my interest and my intention.
'Cultural Values: A Forgotten Strategy for Building Community Support for Protected Areas in Africa', Infield, M. (2001) *Conservation Biology*. 15(3): 800–802.

over their rights, and on the other, that Indigenous People are the conservation movement's closest and most natural allies.

Though most conservation charities today have Indigenous Peoples programmes, to me these look more like rapid responses to pressure than genuine efforts to understand how the links between these peoples and their worlds could help achieve conservation in practice. And it is clear that despite the multitude of individuals, organizations and institutions that accept arguments for values-based approaches to protecting nature, mainstream conservation has remained largely impervious to them.

So ... why *does* mainstream conservation reject the idea that cultural values are the foundation of connections to nature, whether in developed or developing countries, and that more effective and sustainable programmes can be built on these connections? It seems to me that it is the culture of conservation itself that lies at the root of the problem.

Conservation has put its faith in science, made it the touchstone of the profession. The trouble is that most modern conservationists seem unable to expand what they believe to be the absolute and all-embracing truth of science in order to allow other worldviews to stand by its side, on an equal footing. That modern conservation started in the west, even if informed by a very different perspective at the time, seems to have cemented the western ideology at its core: an ideology based on dualism, belief in a single god – and, all too often, the idea that there is a single right way of doing things.

Issues of power and control are also at work, as illustrated by my story of Mburo. Though many working in conservation have shown a willingness to share control, few seem willing to make any compromise relating to the determination of what protected areas actually mean.

Constructing moral justifications for conservation leads to endless arguments, because morality is relative, contextual and cultural. What, then, about sustainability? If we truly accept the idea of it as a guiding principle we are driven to select only the political, social and economic systems that can deliver it. Difficult decisions will remain, however, because, as we have seen, sustainability is itself complex; but balancing what we want for our children may help us avoid narrow and contested political and moral arguments.

Humanity cannot survive if our material needs are not met; yet this alone is not enough for us to flourish. To thrive as individuals, we must fulfil our human and social potential. These are strong drivers of behaviour, because they help us approach completeness. The modern world, with all its material and technological wonders, has succeeded in confusing needs with wants. We suspect we are being short-changed, yet are persuaded that the material life is everything; we don't just want the latest gadget, entertainment or fashion trend, we *need* them. Just as the wrong leaders can send us to war by convincing us that enemies threaten our basic needs, thus stimulating the irresistible instinct for survival, modern life has us believing that things – stuff, possessions – will fill the void opened by the erosion of faith, spirituality and community.[191]

---

191 Wish Shopping, an online shopping site, advertises an endless supply of things for those with 'timeonyourhands', seemingly for nothing more than meeting an immediate wish to entertain themselves and their friends with stuff. https://www.wish.com

The situation is serious. The global enterprise that dominates the world is firmly wedged in the driving seat and has its foot 'stuck on the accelerator'.[192] If allowed, though, nature will provide for our needs and most of our wants, as it always has. And if we nurture it, nature will supply the emotional, spiritual and psychological sustenance that humanity also needs. Reconnecting people to nature and all its gifts will demonstrate the ultimate emptiness of materialism as a philosophy and a way of life. Efforts to conserve nature can lead efforts to fashion new multi-tiered, transformative relations between people and the natural world relevant to the 21st century. Belief in the redemptive power of nature informed the birth of the modern conservation movement in America not so very long ago. I believe this understanding must be reinstated at the centre of conservation, but it must be broadened from its Christian origins to encompass and validate the multitude of beliefs and worldviews that connect people to nature.

Ethical and philosophical arguments for conserving nature for its own sake are valid, and for many are the most important argument – but the wellbeing that nature delivers to humanity is important too, and is argument enough for many. Nature benefits us all when we value what it gives us. When truly acknowledged, the values of conservationists and communities can be sewn together into a single endeavour. Such a partnership would be more powerful and persistent than any that has yet been built around science and economics. Approaching conservation through communities is right; but appealing to them on the basis of their material needs alone is wrong. If modern conservation is to succeed, it must embrace the culture of communities and integrate these connections between people, place and the non-material world into the design of protected areas and conservation practice. The world of nature is and always has been more than the merely physical. If we want to protect nature, all its intrinsic and endowed values need to be placed at the centre of human existence. Then, by our saving nature, nature may save us.

When I reflect on the community conservation programme with the Uganda National Parks which ran from 1990 to 2000, I see successes and disappointments. Mburo was saved from the death by a thousand cuts that had seemed inevitable, and despite ongoing problems its biodiversity, its beauties, its special nature and its values look set to remain. The engagement in new and positive ways with the communities around the park was key to this achievement, and this recognition helped cement community conservation at the heart of the Uganda Wildlife Authority as it grew and evolved. That we in our community

---

192 Nearly ten years ago, Ban Ki Moon, the then Secretary-General of the United Nations, said, 'Our foot is stuck on the accelerator and we are heading towards an abyss.' And we still are. He was referring specifically to climate change and the need to make rapid progress towards reducing carbon emissions, but the picture he presents is relevant to unsustainable economic development generally, driven by the global system of modern capitalism that demands never-ending economic growth.

approach were unable to engage meaningfully with the Bahima despite our concerted efforts guided me towards a recognition of the importance of cultural connections to places and nature and the importance of integrating this into conservation policy, planning and practice.

The experiences and achievements of the cultural values programme between 2005 and 2015 are instructive, too. The ease with which the work progressed in the Rwenzori Mountains compared with Mburo and the forest parks was notable. Not unreasonably, the park managers were interested in practical support for what they were expected to achieve; it was the park management's integration of the values and interests of the local Bakonzo and Baamba people and their leaders that brought immediate and direct support to them. The wardens of Mburo and the forest parks, however, seemed to see a prospect of too little returns for them in accepting that the Enyemibwa or the Batwa had any place in their parks.

In the final analysis, however, it was the institutional culture of the Uganda Wildlife Authority and the values of its managers that determined which of the local values would be recognized and accepted. In Mburo, managing the herds of 'cultural cattle' would have demanded a fundamental change in the way the wardens thought about the park as a protected place. Similarly, at Bwindi it proved too hard for the conservationists to accept that Batwa values should contribute to defining its meaning and to recognize that the Batwa were central to the park's future.

In fairness, the failure of the Uganda Wildlife Authority's willingness to engage with many of the cultural connections that our project had proposed as a positive force for conservation was not a failure generated by the approach itself. The problem in fact had resulted from the reluctance of the authority to engage with communities on a level that might be construed as sharing power. The authority, despite working closely with communities from the 1990s, was unable to go beyond a mere consultation exercise with the community members, and would not engage with them as partners.

This greatly limited the impact of the cultural values approach. The authority would share neither the determination of what the parks represented and meant, nor allow the activities within them that would retain – or restore – the true cultural meaning of the land enclosed within the parks. For Mburo, which has been centre stage of this story from its start, the consequence at the time of writing is that although the Enyemibwa and all their associated values could be part of the park, they remain excluded. In practice, however, as described earlier, on moonless nights the cattle are still pushed by their herders to crop the quiet night pastures of the park, their crescent horns glowing darkly as they reflect the starlight – but all hidden, concealed, denied. And sadly, Mburo remains focused on the values of its biodiversity and the value of its tourism earnings, impervious to the power of history and ancestry and to the connections through which the Beautiful Beasts would illuminate the Beautiful Land.

Recommended background reading

Adams, J. S. and McShane, T. O. (1992). *The Myth of Wild Africa: Conservation without Illusion.* New York, W.W. Norton.

Croll, E. and Parkin, D. (1992). *Bush Base: Forest Farm; Culture Environment and Development.* London and New York, Routledge.

Doornbos, M. R. (1978). *Not All the King's Men: Inequality as a Political Instrument in Ankole, Uganda.* The Hague, Paris, New York, Mouton Publishers.

Hulme, D. and Murphree, M. (eds) (2001). *African Wildlife and Livelihoods; the Promise and Performance of Community Conservation.* Oxford, UK, James Currey.

Infield, M. (2002). *The Culture of Conservation: Exclusive Landscapes, Beautiful Cows and Conflict Over Lake Mburo National Park, Uganda.* Unpublished PhD Thesis, University of East Anglia, UK.

Infield, M. (2003). *The Names of Ankole Cows.* Kampala, Fountain Publishers.

Karugire, S. R. (1971). *A History of the Kingdom of Nkore in Western Uganda to 1896.* London, Oxford University Press.

Kinloch, B. (1972). *The Shamba Raiders: Memories of a Game Warden.* London, Collins and Harvill Press.

Marnham, P. (1981). *Dispatches from Africa.* London, Sphere Books.

Monbiot, G. (1994). *No Man's Land: An Investigative Journey through Kenya and Tanzania.* London, Macmillan.

Morris, H. F. (1964). *The Heroic Recitations of the Bahima of Ankole.* Oxford, Clarendon Press.

Museveni, Y. K. (1997). *Sowing the Mustard Seed: The Struggle for Freedom and Democracy in Uganda.* London, Macmillan.

Nash, R. (1982). *Wilderness and the American Mind.* New Haven, Yale University Press.

Neumann, R. P. (1998). *Imposing Wilderness; Struggles over Livelihood and Nature Preservation in Africa.* Berkeley, University of California Press.

Pakenham, T. (1992). *The Scramble for Africa: The White Man's Conquest of the Dark Continent From 1876 to 1912.* New York, Avon Books.

Schama, S. (1996). *Landscape and Memory.* London, Fontana Press.

Wells, M., Brandon, K. and Hannah, L. (1992). *People and Parks: Linking Protected Area Management with Local Communities.* Washington, World Bank / WWF / USAID.

Williams, D. R. (1987). *Wilderness Lost: The Religious Origins of the American Mind.* London, Associated University Press.